RADIATION AND HEALTH

MELMAN, NELI

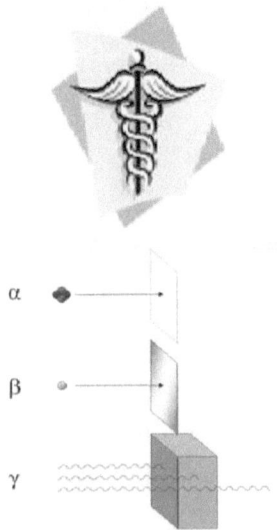

POPULAR MEDICINE

RADIATION AND HEALTH

Washington
2010

Library of Congress Control Number:		2010900549
ISBN:	Hardcover	978-1-4500-2743-4
	Softcover	978-1-4500-2742-7
	E-book	978-1-4500-2744-1

To order additional copies of this book, contact:
Xlibris Corporation
1-888-795-4274
www.Xlibris.com
Orders@Xlibris.com
73924

CONTENTS

IN MEMORY OF THE VICTIMS OF RADIATION

A nuclear accident anywhere is a nuclear accident everywhere
(Aphorism)

The splitting of the atom has changed everything except our way of thinking and thus we drift towards an unparalleled catastrophe.

—Albert Einstein

Acknowledgments

I thank Dr. D. Branovan and his colleagues for their help in translating the book from Russian into English

INTRODUCTION

The latter stages of the 19th century and decades that followed were marked by a number of great breakthroughs. Among these, a prominent place belonged to the discovery of X-rays and nuclear fission. These advances resulted in arrival of otherwise unforeseen changes that affected practically all aspects of mankind, including the health of people.

The detection of X-rays was a fortunate discovery made by a genius mind of German scientist **Wilhelm Conrad Roentgen** (on November 8th, 1895). Subsequent years that subsequently followed were devoted to the study of their properties. It was established that X-rays, also called Roentgen rays in honor of their inventor, possess a number of unique attributes, such as ability to ionize air, reflect from a surface, and penetrate matter. Such properties enabled a wide scientific and practical application, especially in the medical arena.

The discovery of nuclear fission evolved from intensive long-term research by outstanding scientists in different countries. As a result, approximately 30 physicists, chemists, and biologists received the Nobel Prize for their distinguished work. Several accomplishments in particular stand out and include the discovery of spontaneous radioactivity (**Antoine Henri Becquerel, Pierre Curie, and Marie Sklodowska-Curie, 1903**), synthesis of new radioactive elements (**Frederic Joliot and Irena Joliot-Curie, 1935**), discovery of novel radioactive compounds from irradiation of neutrons (**Henrico Fermi, 1938**), and invention of a cyclotron (Ernest Lawrence, 1939). The first nuclear reactor in the world was designed by the Nobel Prize laureate Henrico Fermi in 1942. The first woman to win the Nobel Prize in the field of medicine was **Rosalyn Yalow,** who applied radioactive substances for diagnostic purposes in a number of diseases (1977). Considerable recognition belongs to a number of other Nobel Prize laureates and distinguished scientists who advanced the field of radiation research.

It is important to appreciate the significance of radiation research in clarifying our understanding of radiation. Over the years, significant discoveries about radiation and its associated hazards have been widely recognized with high national as well as international awards, degrees, and ranks. Complexity and diversity of problems associated with radiation has resulted in creation of novel scientific and applied specialties, such as nuclear medicine, radiobiology, radio veterinary science, radio physics, radiochemistry and many others.

The decision to investigate numerous facets of radiation and its impact on people largely stemmed from two tragic events of the past century, both of which loudly resonated on the international stage:

* nuclear bombardment of Japanese cities Hiroshima and Nagasaki (August, 1945);
* nuclear power plant accident in Chernobyl (April, 1986, Ukraine, former Soviet Union).

The escalation of nuclear arms—a process that began in the 1940-50s and persisted until this day—remains very dangerous for the mankind. Unfortunately, there is no assurance that the use of nuclear weapons can be avoided in the future. Moreover, the accruing threat of radiation-based terrorism, which dates back to the 1980s, presents significant concern and a sense of urgency to examine various aspects of this emerging problem.

So what are some of the basic considerations in research dealing with the effects of radiation on people? Firstly, it is important to note that results of radiation research are not always unequivocal. To a certain extent, this stems from constant changes in methodology and techniques, as well as from several other factors.

For almost a century, research analyzing the impact of radiation on people focused on the following issues:

* Investigation of radiation's mechanisms of action.
* Determination of the relationship between induced injuries and various characteristics of radiation (nature of the source, dose, the duration of exposure, etc.).
* Study of detrimental effects of radiation on different biological levels (organ, cellular, sub-cellular).
* Improvement in the diagnosis of radiation injuries.
* New methods and means of treatment.
* Primary and secondary prophylaxis.
* Medical and social rehabilitation of victims.
* Methodology for long-term surveillance of victims and their offspring.

Research directions listed above have yielded positive outcomes to this date. However, despite all these accomplishments, numerous issues remain unresolved and are subject to further urgent and thorough studies.

Cooperation and integration of various national and international agencies carries paramount significance in training of specialty services, preparation of specialists at all levels, and improving the scientific literacy of the populations. Particularly in the U.S., substantial government grants as well as private donations are allocated to address such important objectives.

Since the beginning of the 21st century, with combined participation of specialists with expertise in various backgrounds (radio physicists, radiobiologists, mathematicians, physicians, etc.), the following areas have also been under intense investigation:

* Bio dissymmetry and creation of new automated diagnostic systems.
* Novel protective equipment and additional methods of prophylaxis and treatment (antibiotics, etc) during radiation exposure.
* Safety the sources of artificial radiation.
* Increased research on the damaging effect of sources of natural radiation (cosmic rays, etc.).

Chernobyl's catastrophe opened a new chapter in nuclear medicine—the study of complications arising from exposure to small doses of radiation. This very complex and critical issue represents—and will continue to do so for a long time—one of the central topics of research investigating damaging effects of radiation in people. Even prior to Chernobyl's fallout, in the nineteen-fifties a distinguished Russian scientist **A. Sakharov** expressed concern that small radiation levels may exert detrimental effects on the human organism.

Long-term experiences with radiation attest to the fact that hopes of the mankind for a "peaceful atom" have fallen short of expectations. Humanity's experience with radiation parallels many other scientific and technological advances, in that any significant progress may be accompanied by serious failures and catastrophes. This, in turn, may negatively affect the health of people. Therefore, safety measures for artificial, or man-made, radiation represent a matter of critical significance.

Analysis and generalization of results from ongoing radiation research take place on a regular basis at numerous local, national and international meetings, during which practical recommendations and decisions on further research are made.

It is important to note that a number of scientific recommendations, in particular in Ukraine, Belarus and Russia, do not find support in government establishments, resulting in various forms of protest, such as demonstrations, strikes, hunger strikes, etc.

Currently, scientific publications addressing the topic of radiation number in thousands. Issues pertaining to radiation's impact on people continue to generate concern in the world community. They are reflected in fiction, art, theatrical productions, and films, with many of such works receiving high marks of achievement.

Unfortunately, the number of publications addressing the impact of radiation on population's health remains limited. For a variety of reasons, these publications are not always easily accessible to a broad audience of readers, especially those that are victims of radiation exposure from various sources (accidents, radioactive waste products, etc.).

This book presents an in-depth analysis of various authoritative materials from global literature and draws from a personal experience in the area of nuclear medicine.

The word "radiation" is included in people's minds as the image chug something danger. This is reflected in literature, art, films. People does not have sufficient opportunity to learn the basics of radiation medicine. This problems are usually absent in the preparation of doctors and middle medical workers.

Dear Readers, If you're unfamiliar with any word, please refer to the "Glossary of terms" at the end of the book.

GENERAL REPPRESENTATION ABOUT RADIATION

Nuclear Fission (Splitting of the Atom)

All matter in the world consists of small particles called *molecules*, which function as a unit and consist of two or more atoms. *Atoms* participate in chemical reactions and may further split into smaller constituents, despite their name's derivation from a Greek word "indivisible." Molecules and atoms are not visible to the naked eye. The size of an atom equals to one-billionth of a centimeter. The 'period mark' at the end of this sentence can fit 100 billion atoms.

A positively charged nucleus occupies the center of the atom and determines its properties. The nucleus consists of protons and neutrons. Particles called electrons spin around the nucleus in a well-defined set of orbits. The number of electrons in the atom equals to the number of protons in the nucleus.

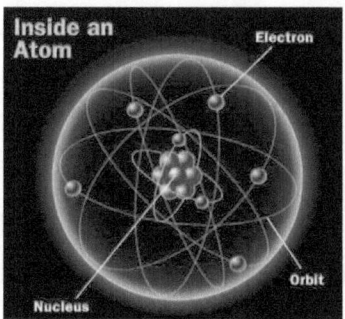

Structure of atom (Scheme)

The majority of atoms comprising various compounds found in the environment are stable, i.e. they do not undergo chemical breakdown that may potentially result in radiation.

In contrast, substances containing unstable atoms also exist in nature, for example uranium. They spontaneously break apart to become other elements and in this process release radiation energy (alpha, beta, and gamma). Such substances are commonly referred to as radioisotopes. The quantity of naturally existing radioisotopes is much less than of those artificially created (200 and 400 names respectively). In the recent decades, technically advanced nuclear reactors have been developed to artificially split atoms. Design and size of these reactors largely depends on a specific task at hand (research, medical, energy source, etc.).

During nuclear fission, i.e. division of the atom, large quantities of energy are released. In this process, an emission of radiation takes place (the name comes from the Latin word 'radius', which means a 'beam'). Atomic nucleus of cells of human tissue maintenance in chromosome. This double helix of DNA with a set of gene-carriers of genetic (hereditary) code.

Following several cycles of disintegration, the total number of radioactive nuclei decreases. The time it takes to reduce the quantity of these elements in half defines a half-life period. It is important to emphasize that large variations in half-life exist among various substances—from several days up to hundreds and thousands of years (**Table 1**).

Radionuclide	Half-life	Radiation Type
Americium[241]	432.2 years	Alpha, Gamma weak
Cesium[137]	30 years	Beta, Gamma
Cesium[134]	2 years	Gamma
Iodine[131]	8 days	Beta
Iodine[129]	1.5 million years	Beta
Iridium	74 days	Beta, Gamma
Plutonium[238]	24.000 years	Gamma
Strontium[90]	29 years	Gamma
Strontium[89]	50 days	Gamma

Table1. Half-life and radiation type of several radionuclides.

Isotopes of the same chemical element may vary based on duration of their half-life period (for example, iodine and strontium). They are often referred to as nuclides.

A famous British physicist **Ernest Rutherford** was first to propose "alpha" and "beta" rays. X-ray close to γ—ray.

Alpha (α)—rays are generated during the decay of uranium, thorium, radium and plutonium. They carry a large quantity of energy; however, their ionizing power is not very strong. These rays fail to penetrate even a sheet of paper and practically do not enter the organism through intact skin and mucous membranes. However, entry into the organism remains possible through open wounds, including via inhaled air and consumed food. Under these circumstances, alpha rays may pose danger to the health and even a life of an organism.

Beta (β)—rays arise primarily as a result of radioactive decay and travel at very high velocities. They are dangerous when deposited onto skin surface as well as after gaining entry into an organism. Penetrating power of beta rays is much stronger than that of alpha rays—they pass through tissues of an organism to the depth of 1-2 cm.

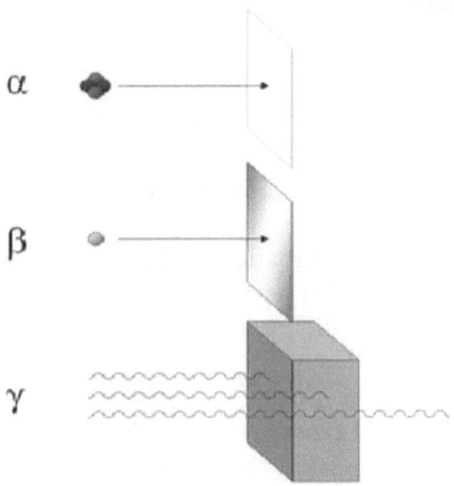

Types of rays and their penetration

Gamma (γ)—rays have the shortest wavelength, yet possess the strongest penetrating power. They travel at the speed of light. Thick plates of lead or concrete can protect gamma ray penetration into the body. Gamma rays are similar to X-rays in terms of their properties.

Alpha-, beta-, gamma—rays differ release a certain amount of energy, varying penetration. Their effects on the body varied. Penetration of these rays is inversely proportional to the length of them. Shorter rays more ability penetration then longer.

Basic characteristics of radiation

Radiation possesses a whole series of unique characteristics. They can be divided into *quantitative* (X-ray, Gray and others) and
qualitative reflecting the impact, particularly on the human

The use of any other unit determined by the object of study (people, environment and others).

Quantitative characteristics of radiation

Quantitative assessment of the dose describes either the absorbed dose or its power. International units have been adopted for quantitative assessment of radiation:

Curie (CI): International traditional unit of radiation measurement, equal to radioactivity of 1 gram of pure radium. It reflects the decay rate per second of a radioactive substance. It is named in honor of Marie Sklodowska—Curie.

Becquerel (BQ): standard international unit of radioactivity, equal to one decay per second. In essence, this represents the same unit as Curie, only in the international system of units (SI). It is named in honor of a famous researcher Antoine-Henri Becquerel.

RAD (Radiation Absorbed Dosage): a unit of energy absorbed from ionizing radiation, defined as accumulation of 100 ergs per 1 kilogram of irradiated material.

Gray (GY): a unit of absorbed radiation in the "SI" system, equivalent to 100 RAD.

REM (Roentgen Equivalent): a unit of measurement describing damaging effects of radiation, usually expressed in Severs.

Severe (SV): the amount of ionizing radiation required to generate the same biological effect as one RAD of x-rays, equivalent to 100 REM.

BER (Biological Equivalent of Roentgen): a unit of measurement of equivalent biologic radiation dosage.

Roentgen: a unit of radiation exposure.

Conversion of radiation units:

REM=dose in RAD multiplied by a coefficient for alpha, beta, and gamma radiation (1, 20, and 10 respectively).
REM=100 x Sever (SV).
Sever x100= REM
Micro-SV=one thousand of SV.
Gray=100 RAD.
RAD=0.01 Gray.
BER=100 x SV.
BER=Roentgen.
Becquerel (BQ)= 7 Curie (CI)
1 GY=1 SV = 100 RAD= 100 REM

It is important to note that terms used for radiation measurement undergo modification and new ones are proposed periodically. At present, measurement units of Sever and Gray are employed most frequently for determining the amount of radiation exposure. Furthermore, various quantitative indices are widely used for establishing the degree of the radiation content in food, water, soil, building materials, raw materials, and industrial products. The use of the same radiation units facilitates efficient communication and understanding among researchers. Besides measurement indices listed above, the impact of radiation may also be evaluated using many other parameters (the name and number of the isotope, distance from the source, duration of exposure, etc.).

Determination of the radiation dose

Radiation does not possess color or smell—therefore, its detection is only possible with the aid of special instruments called radiometers (radiation dose monitors, dosimeters). Of critical value is their ability to accurately determine the level of radioactivity in various surroundings (soil, air, water, etc.) as well as in living organisms.

The field within physics dedicated to determining the quantity, intensity and distribution of ionizing radiation in space and time is called dissymmetry. In other words, dosimeters represents measurement of energy, which is transmitted to a particular body of mass by radiation. Therefore, a radiation dose reflects the amount of absorbed energy. Measurement of radiation levels in human organisms includes not only the determination of general radioactivity, but also measures of intensity in various tissues and organs (thyroid gland, bone, etc.).

Devices measuring radiation are called radiometers or dosimeters

Various kinds of dosimeters

Direct measurement of radiation dose absorbed by living tissues is not possible. Hence, indirect measurements of the dose are performed using specific physical or chemical media. They are referred to as dosimeters detectors, and consist of liquid, gaseous, or solid material. Measurement of the radiation dose transmitted to a certain medium is calculated from its exerted effects on the atomic level.

The following types of detectors exist—ionizing, semiconductor, luminescent, thermal and chemical. Various detectors are characterized by their degree of sensitivity to specific radioactive materials, and this determines, the use of a particular detector depends on the composition of radiation (X-rays and gamma rays, or alpha and beta particles) and its intensity.

Dosimeters are manufactured for individual use, scientific research, and military purposes. They may be designed as stationary or portable units. The purpose for their utilization determines size, form and construction of a particular dosimeter. Three types of dosimeters exist for measurement of beta and gamma radiation, detecting low or background, medium, and high ranges of intensity. Specialists called dosimetrists perform the work related to radiation measurement. Improvement of dosimeters represents an important challenge in radiobiology and nuclear medicine. Various specialists dedicate their work to ongoing improvement of such devices.

Unfortunately, it is not always possible to determine the exact level of individual as well as public radiation exposure. Yet, such information represents a critical prerequisite not only for designing rational therapeutic and prophylactic measures, but also for subsequent prognosis of the induced injury. Hence, during the past five decades, various methods of retrospective estimation of accumulated radiation dose have been developed. Japanese scientists, mainly on the basis of special physical and mathematical calculations, developed a system called DS86. With the help of this system, radiation exposure has been estimated in survivors of nuclear bombings of Hiroshima and Nagasaki.

Newer methods of physical and biological dissymmetry, designed to retrospectively calculate accumulated radiation levels, have been recently proposed and developed. Such methods are generally referred to as reconstructive. Spectroscopy of the electronic paramagnetic resonance of tooth enamel represents one of these promising methods EPR—dosimetry. This approach is especially useful for radiation victims of Chernobyl' catastrophe. In the opinion of certain specialists, this technique represents the most sensitive and specific method of estimating the ionizing radiation dose. Of significance, EPR dosimetry is not limited by the time that elapses following radiation exposure.

Estimation of absorbed radiation from plutonium and strontium is available in specialty laboratories, which may calculate the radiation dose from urine, feces, or breast milk samples. Scientists in Ukraine recently proposed a new method of retrospective evaluation of radiation exposure by directly examining the surrounding matter, such as soil, plants, etc.

The International Program on the Health Effects of the Chernobyl Accident (IPHECA) recently proposed a model to estimate radiation levels delivered by radioactive iodine to the thyroid gland. This would be most applicable to those geographic locations where direct measurements of radiation were not performed.

Used cytogenetic method (FISH) to retrospectively dose in acute and chronic exposure to radiation. It allows you to predict the adverse effects induced by radiation

An American scientist developed a simple method of blood analysis, which allows to determine the amount of radiation that has been absorbed by the body. This method needs further approval.

Studies in this area continue to evolve, with new methods steadily replacing old techniques.

Qualitative characteristics of radiation

Qualitative characteristics of radiation are used depending on the specific situations of radiation contamination.

Dose of the ionizing radiation—amount of energy, absorbed by a unit of mass in the environment.

Absorbed dose—quantity of energy, absorbed by a unit of mass of the irradiated body.

Equivalent dose—biological activity of different forms of radiation (alpha, beta, gamma, etc.). The equivalent dose is equal to the product of the absorbed dose of a given type with the corresponding quality coefficient. For X-, gamma, and beta rays, the coefficient equals to 1, for alpha rays -10, and for neutron emission -20. It is usually expressed in Severs (SV or Biological Equivalents of Radiation (BER).

Dose Power—value allowing to make prognostic inferences with regard to the level of radioactive exposure. In each specific instance, this value is equivalent to a dose of radiation received per unit of time (second, hour, etc.).

Effective equivalent dose—reflects the aggregated effect of radiation on an organism. This value is summated by applying weighted coefficients to various organs and tissues of the body, corresponding to their degree of sensitivity to the effects of radiation. In locations with a radioactive source, such as in Chernobyl following the accident on the nuclear power plant (April, 1983), an annual effective dose may be determined instead.

Specific activity—parameter describing the level of radioactive contamination in food, water, soil, building materials, raw materials and industrial products. The use of specific detector types in different circumstances. Consequently, qualitative characteristics of radiation allow to receive representation about chapter of its influence on the human and environment.

Sources of Radiation

Radiation sources fall into two major categories—**natural** and **artificial,** with the latter forming a significantly larger group.

Sources of natural radiation is less than artificial.

Sources of Natural Background Radiation

Natural background radiation has existed since the beginning of the universe, as a byproduct of the energy-releasing chemical reaction involving decomposition of atomic nuclei. This process, known as nuclear fission, continues to occur ubiquitously in the solar It is its quality and is referred to as "natural," "background" radiation. In Russian literature often uses the term "normal radiation." system.

Recent scientific data indicated, the term is incompetent, because natural radiation, under certain conditions can have a damaging effect on organisms.

Life on Earth has evolved in a constant presence of natural background radiation. Radiation truly represents a basic element of the environment, just like the solar rays from the sun and stars, temperature, pressure, and time. Natural background radiation arises from the following three sources: **cosmic radiation** (from high-energy protons, electrons, X—and gamma rays that continuously strike. Earth (as it moves through space), **terrestrial radiation** (from radioactive materials found throughout nature, such as uranium, thorium, potassium, radium, and their radioactive decay products), and **internal radiation** (from radioactive potassium^{40}, carbon^{14}, lead^{210}, and other isotopes present inside human bodies from birth).

The relative contribution from these diverse sources varies. Radon, a naturally occurring gas formed from the radioactive decay of uranium^{238} in rock and soil, accounts for 55% of the total radiation dose that people receive each year. Other naturally occurring radioactive materials found in Earth's crust, as well as cosmic radiation, contribute about 8% each to the annual radiation exposure. Isotopes located inside human organisms contribute an additional 11%. These radioactive materials enter the organism through food and air and distribute in tissues primarily as potassium and carbon radioisotopes. Specific contribution of known radioisotopes (4%).

Cosmic radiation is formed in the solar system by the sun and stars, and earth's atmosphere absorbs a significant proportion of this radiation. This secondarily creates an additional source of steady radiation release for the planet. Not surprisingly, the dose of penetrating cosmic radiation varies in different geographic regions of the globe, mainly affected by the differences in elevation and the strength of local magnetic fields.

According to the data from the American National Council on Radiation Protection and Measurements, radiation exposure from cosmic rays at sea level amounted to 40 µREM per year during the time period from 1945 to 1995. During the same time frame, the annual dose from radioactive materials in rock and soil was 55 µREM.

Cosmic radiation in the atmosphere is especially dangerous for cosmonaut, pilots and, to a lesser degree, passengers traveling in aircrafts. For every hour while airborne, the individual radiation dose reaches 0.5 µREM. This issue has generated considerable attention in the U.S., Canada, and countries of the European Union. In response, exposure limits have been established for both the crew and the passengers.

Radiation content of air, water, and food varies widely, ranging from 20 to 400 µREM annually. Smoking one pack of cigarettes every day for one year contributes an additional 40 µREM of radioactive exposure.

Natural background radiation registered in each state of the U.S. is not uniform. Naturally occurring radiation sources primarily determine the individual exposure levels for people in the U.S. Artificial radiation sources contribute only a small percentage toward annual exposure (**Table 2**).

Natural Sources

Source	Dose in µREM/year	Dose in µSV/year	% of Total
Radon	200.0	2.0	55%
Other Radio nuclides	27.0	0.27	8%
Cosmic Rays	28.0	0.28	8%
Internal	39.0	0.39	11%
Sub-Total	300.0	3.0	82%

Artificial Sources

Source	Dose in µREM/year	Dose in µSV/year	% of Total
Medical X-rays	39.0	.39	11%
Radiation Therapy	14.0	0.14	4%
Consumer Products	10.0	0.1	<0.03
Sub-Total	6.3	0.63	18%
Grand Total	306.3	3.06	

Table 2. Annual exposure levels from various natural and artificial radiation sources (%/ year).

As such, under normal conditions in the U.S., natural radiation sources are 4.5 times higher than artificial. An unusually high background radiation (0.5-1.2 REM per year) has been reported in several places around the globe—Shri Lanka, India, Brazil, Iran, and Italy. Evidently, this is associated with a large bedding of uranium ores and the outcrop of radon sources. Such elevated background radiation levels form a risk factor for lung cancer in those geographic locations. Furthermore, uranium and its derivatives (thorium, radon, etc.) are particularly dangerous for the workers inside the mines, especially when preventive measures are not thoroughly instituted and followed.

According to the data from Russian researchers, 90% of the global population annually receives radiation levels equivalent to 0.3-0.6 µSV. Roughly 3% are exposed to approximately 1 µSV, while 1.5% acquire above 1.4 µSV.

Through evolution of thousands of generations of living organisms, practically all species have adapted to exist under conditions of persistent background radiation. Although their quantity is small, some radioactive materials may reside inside an organism throughout a lifetime. To counteract this long-term exposure, organisms have developed biologic mechanisms designed to neutralize negative effects of radiation.

In recent decades, exposure levels from natural background radiation have increased in many places across the planet. This is directly linked to a widespread use of radioactive materials in numerous branches of industry, power engineering, medicine, and military. Several exposure limits have been established by overseeing nuclear agencies. For instance, the permissible dose for people working with radioactive materials has been set at 5 µREM per year. Background radiation in a radius of 50 miles around nuclear power plants should not exceed 0.01 µREM. In general, background radiation is routinely monitored in the vicinity of nuclear power plants, cyclotrons, nuclear weapon test sites, nuclear submarines, and sites of radioactive waste disposal.

Until 1992, the annual exposure level in the U.S. from background radiation constituted 300 μREM, and during subsequent years rose to 360 μREM. As previously discussed, this background radiation derives from natural (approximately 80%) as well as artificial sources (20%). Reduction of background radiation remains at the center of numerous ongoing research efforts. The significance of this issue has been in turn supported by recent scientific data suggesting direct detrimental effects of even small radiation doses on human health.

Sources of Artificial Radiation

Artificial radiation is similarly generated by nuclear fission—a chemical reaction of atomic decay, yielding a concomitant release of radiation energy. Sources of artificial radiation may exert an incomparably greater negative impact on people by releasing much larger doses of radiation. Furthermore, artificially created radioisotopes are characterized by longer half-lives, greater tendency for spontaneous decay, and a stronger penetrating ability—all of which significantly add to the potency of artificial radiation.

Following the invention of a cyclotron in 1935, further scientific research on nuclear fission continued in the Moscow area of Obninsk (former Soviet Union) and in California at the University of Berkeley (Ernest Lawrence and colleagues). This resulted in a steady increase of a number of different methods and techniques designed to generate artificial radiation. Concomitantly, considerable progress was made in improving the overall quality and efficiency of power plants as well as methods of personnel protection.

Existing sources of artificial radiation, capable of exerting detrimental effects on the health of people are presented and discussed in further detail below.

Radiation and Medicine

General formation

The discovery of X-rays and nuclear fission (splitting of the atom) served as a strong stimulus to evaluate their role in human health. Nuclear medicine evolved as a separate field within medical sciences specifically to address this emerging issue. Its history has been marked not only by keen observations chronicling development of tumors in a setting of radiation exposure, but also by extensive scientific studies. In the early stages, the field of nuclear medicine exclusively investigated radiation effects on people. Over time, the scope of nuclear medicine expanded to employ radiation in the diagnosis and treatment of various diseases.

Radiation oncologists routinely collaborate with nuclear physicists, chemists, biologists (those studying radiation effects in animals), and ecologists (those evaluating radiation's impact on environment). During the past century, radiation-based diagnostic and therapeutic methods have continued to increase and improve. It is difficult to imagine the practice of medicine today without the use of these methods in the diagnosis and treatment of common as well as rare illnesses. Today, radiation technology is extensively applied in cardiology, pulmonologist, gastroenterology, neurology, obstetrics and gynecology, urology, nephrology, orthopedics, etc. In essence, every medical field has adopted some use of radiation.

The pioneering discoveries of Pierre Curie and Marie Sklodowska-Curie allowed to seek practical applications of radioactive isotopes in medicine.

The first documented use of radiation took place in France in 1901, when a radioisotope was employed to treat a cancer patient. Three years later, a dermatologist from Melbourne (Australia) also used a radioactive isotope in a patient with skin cancer. After 1924, radioactive products of radium were routinely administered intravenously for diagnostic purposes. Nuclear medicine began to develop successfully in the 1950's. Its foundation was solidified by efforts of the first woman to win the Nobel Prize in medicine—Dr. **Rosalyn Yalow**—who used radioactive iodine (Iodine[131]) to diagnose diseases of the thyroid gland.

There are reports of the death of 336 people who participated in the various studies of radiation in the early 20 century. It is know cases of cancer in the medical staff assisting the victims of the Chernobyl accident.

Presently, gamma rays emitted from decay of short-lived radioisotopes (iodine, bismuth, cobalt and others) are preferentially utilized for medical purposes. These radioactive compounds are generated in specialized cyclotrons and can be administered to patients intravenously, by inhalation, and through the gastrointestinal tract.

Methods of Nuclear Diagnostic

Radiation-based diagnostic techniques may be used in a variety of settings. Some examples include detection of earlier stages of cancer and metastatic spread as well as assessment of blood flow in the vessels of the heart, kidneys and other organs. Nuclear studies are also employed during procedures designed to estimate the function of the heart, lungs, the liver, kidneys, thyroid gland, etc. Presently, radiation-based diagnostic modalities comprise1.9% of all diagnostic procedures in developed countries.

Study Type	Radiation Exposure (µSV)
X-Ray of extremities	0.01
X-Ray of teeth	0.04-0.15
X-Ray of the head and neck	0.20
X-Ray of the chest	0.06
X-Ray of the pelvis	0.65
X-Ray of the neck	0.20
Mammography	0.70
Contrast study	4.05
Nuclear scan of the thyroid gland	14.0

Table 3. Radiation exposure levels from various diagnostic methods.

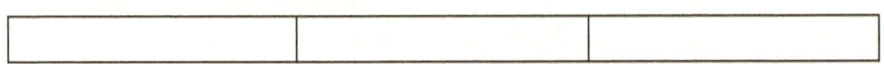

It should be noted that these figures are largely determined by the quality of the equipment used. The more sophisticated apparatus, the lower the radiation exposure.

Interestingly, comparison radio diagnostic beam loaded with natural radiation background. For instant correlation with Natural Background Radiation: computer tomography of the abdomen 3 years, X-ray of bowels 16 months; X-ray of the stomach 8 months; mammography 3 months.

It is useful to express the radiation dose from different radiologic studies in terms of daily levels of natural background radiation.

As previously stated, radioisotopes with short half-life periods are preferentially utilized today, taking advantage of their rapid decomposition. In addition, radioisotopes have a tendency to accumulate in different concentrations across various organs of the body. As a result, this property allows selective assessment of different organs. For instance, Iodine[131] shows predilection for the thyroid gland and, therefore, represents the preferred radioisotope for this organ.

The number of diagnostic procedures in nuclear medicine continues to increase, with a concomitant reduction in emitted radiation. Radioactive pharmaceuticals, such as Technetium Tc-99m, are especially promising and are commonly used as chemical tracers for the diagnosis and treatment of several diseases. Other diagnostic tracers, carrying a comparatively light radioactive dose, are currently in the process of approval for medical use. If necessary, certain radioactive pharmaceuticals may also be used during pregnancy.

Discovery of radioactive tracers represents a great achievement of contemporary science and medical practice. Unfortunately, their widespread use remains inaccessible to many economically underdeveloped countries.

Radiation Therapy

The net exposure to radiation during treatment of diseases is much higher when compared to diagnostic modalities. Annually, over 18 million patients in the world undergo radiation therapy. This constitutes approximately 10% of the total number of all other treatments performed for numerous severe and widespread illnesses Physicians conducting radiation treatment—radiation oncologists—are closely monitored for exposure levels, in order to prevent long-term side effects from radiation.

At present, about 10 different methods of radiation treatment have been developed (external beam, intra operative, stereotactic, etc). Depending on the anatomic location and microscopic structure of the tumor, radioactive materials are introduced via different techniques (external, intravenous, etc). Radiation treatments may be applied to tumors of different tissue types (solid tumors, blood-based cancers, etc). Radiation therapy has been used with therapeutic success to treat cancers of the brain, breast, uterus, larynx, lungs, pancreas, prostate, skin, spine, stomach, soft tissues, leukemia, and lymphoma. Radiation therapy may be combined with other forms of cancer treatment, such as surgery and chemotherapy, to achieve a better therapeutic response. Furthermore, radiation treatment may be used to reduce and control pain associated with cancer metastases or arising from other underlying conditions.

Targeted cancer therapy holds promise in reducing the exposure to the rest of the body by employing tissue-specific radioactive pharmaceuticals. These isotopes allow the delivery of medication directly into beds of cancerous cells, thereby sparing healthy tissues of the body.

Unfortunately, side effects of radiation therapy continue to present a significant challenge. Many complications of radiation exposure (most commonly affecting hematologic, gastrointestinal, and nervous systems) may be either symptomatically treated or even reversed. However, certain side effects may be significant enough to endanger not only the health, but also the life of a patient. This presents a key issue for radiation oncologists as well as other specialists (immunologists, hematologists, etc). In recent years, improvements have been made in the prophylaxis and treatment of complications associated with radiation therapy. Nevertheless, a relatively high efficacy of radiation treatment and low complexity of its application demand further advancements in the management and prevention of associated side effects.

In the U.S the Nuclear Regulatory Commission (NRC) strictly enforces limits for the annual levels of radiation exposure. Above natural background levels, the NRC requires that its licensees limit maximum radiation exposure to individual members of the public to 100 µREM (1µSV) per year, and limit occupational radiation exposure to adults working with radioactive materials to 5,000 µREM (50 µSV) per year.

Occupational sources of radiation

The term occupational defines radiation exposure occurring in the workplace. Today, radioactive materials are routinely utilized within numerous industries, including automotive, aircraft, construction, electrical engineering, oil and gas, and many others. Application of radioactive products in the workplace continues to expand. Specific symbols are used in order to indicate various radiation levels emitted by radioactive substances.

The Symbols of different level radiation

In 2005 World organizations introduced a new character. Despite criticism of the symbol adopted.

The new radiation symbol

International organizations for the safety from radioactive contamination (1998) designate three categories of occupational radiation sources, which vary based on the level of potential hazard:

Category1. **Irradiating installations and Brach therapy.**
Category2. **Stationary industrial sensors with highly active sources.**
Category3. **Stationary industrial sensors with low-active sources.**

Programs for occupational protection and safety must take into account such classifications during their establishment. Furthermore, routine occupational personnel monitoring is necessary for all branches of industry, given a continuing expansion and incorporation of different radiation-based technologies. But we should not forget about the possibility of its negative impact. Atoms for Peace is not always peaceful. One should not forget the possibility of its negative impact.

Nuclear power plants

General Information

Nuclear fission created a new source of relatively inexpensive, yet very powerful electrical as well as thermal energy. As a result, nuclear power engineering began to expand rapidly throughout the world. Regretfully, this concomitantly created serious consequences for the humanity and the surrounding environment. The emergence of nuclear power plants exposed the other side of the so-called "peaceful atom".

Even in the beginning of the twentieth century, as the radiation era commenced, G. Curie voiced his fears regarding its future: "*I am most concerned about who will use the discovery which I made.*" Subsequent developments echoed his concerns as the direst consequences arose from erroneous applications and executions of nuclear scientific discoveries.

Numerous regional, national and international organizations have been established to monitor nuclear power plants, while developing measures for prevention and treatment of radiation exposures. According to the data from the International Atomic Energy Agency (IAEA), currently 31 countries possess a total of 493 nuclear reactors. The predicted energy crisis in the world is forcing numerous countries to continually increase their nuclear energy sources.

Following the accident on the Three Mile Island Init 2 (TMI-2) nuclear power plant in Middletown (Pennsylvania) on March 28, 1979, the construction of power plants in the U.S. was slowed, despite the absence of an official ban. In recent times, however, this issue received renewed attention, with the support for additional nuclear power plants given by the U.S. president George Bush in his speech during the Small Business Administration conference in Washington, in 2005. Now this position change.

It is important to emphasize that nuclear power plants pose danger not only due to their potential for an accident, but for other significant reasons as well. First, nuclear power plants may secretly function as developmental sites for nuclear weapons. Second, they represent attractive targets for embezzlement of radioactive materials for purposes of terrorism. These critical issues form a new focus of attention in a number of countries, including the United States.

According to the data from the International Radiation Protection Association (May 2004), from 1950 to 2001, about 500 registered accidents occurred on nuclear power plants, injuring approximately two thousand people. During the same time frame, fifteen industrial accidents also took place, injuring a total of 30 people. A total of 435 nuclear power plants exist in the world, providing a significant energy source to 32 countries. Nuclear power plants are especially numerous in India, France, Belgium, North Korea, Switzerland, Sweden, Japan, Russia, and Ukraine. They supply approximately 33% to 76% of all electric power to those nations. In the United States, nuclear power plants provide approximately 20% of the total electric supply.

The magnitude of nuclear power plant accidents varies from small radioactive emissions to severe catastrophes. From 1951 to 1986, twelve nuclear accidents were registered in the United States. In the former Soviet Union, from 1957 to 1985, information exists regarding twelve accidents on nuclear power plants (Beloyarsk, Leningrad, etc.), including a small mishap on the Chernobyl's plant. Experts in the field believe that these numbers are largely understated.

A scale was developed to estimate a hazard level generated by significant accidents on nuclear power plants. According to this scale, Chernobyl' accident was rated at seven marks, the Three Mile Island at five, Chelyabinsk at six, and Tomsk at four. This particular scale is not applicable for nuclear weapons.

Especially dangerous is developed in Russia, the so-called floating nuclear power plant (PAES). Started their introduction in some south-eastern countries. According to experts PAES are great danger, because can be widely used by terrorists.

Outcomes of nuclear catastrophes, especially their subsequent impact on human health, depend not only on the extent of the accident, but also on a timely and honest response by various organizations and task forces, such as those administering emergency medical care. This point was vividly illustrated by the accident on the Chernobyl's nuclear power plant. What took place in Chernobyl could be best described as a grave catastrophe, not just an accident. It became one of the biggest technologic disasters of all time, and caused the largest ecological calamity of the departed century.

Limited malfunctions may also affect nuclear power plants and, similar to large accidents, may claim service personnel as its initial victims. How fully such incidents become registered is not very clear. In the former Soviet Union, such information was tightly protected under "seven seals."

Three Mile Island Nuclear Accident

This accident took place on March 28th, 1979, in the second reactor of the nuclear power plant located on the Three Mile Island near Middletown (Pennsylvania). A combination of personnel error and equipment malfunction caused a partial meltdown of the reactor core, resulting in a relatively insignificant leakage of radioactive materials into the atmosphere. This was the most serious accident in the history of the U.S. nuclear industry, even though it did not cause any injuries to plant workers or people of the surrounding community.

Three Mile Island Nuclear power plant

Immediately following this accident, several critical issues were investigated. They included:

* Measurement of radiation exposure levels and a collective radiation dose.
* Estimation of the number of induced cancer cases among people living in a radius of 50 km from the plant.
* Estimation of the number of other fatal as well as non-fatal associated illnesses.

The damaged nuclear reactor was immediately shut down and subsequently repaired. Simultaneously, intensive cleanup of the affected territory was initiated. This area was completely decontaminated of radioactive materials by 1993. A permission to operate the repaired reactor was obtained soon thereafter. Today, the TMI-2 reactor is permanently shut down and de-fueled. The owner says it will keep the facility in long-term, monitored storage until the operating license for the TMI-1 plant expires at which time both plants will be decommissioned.

After the accident, the governor of Pennsylvania, following a consultation with prominent nuclear physicists, recommended that pregnant women as well as preschoolers residing in a radius of 50 kilometers from the reactor were relocated into safer areas. All events associated with this accident were truthfully and extensively reported through various media sources (television, radio, newspapers, etc.).

For the past 28 years, all workers of the nuclear plant (approximately 1,000 people) as well as inhabitants in the 50-kilometer zone have remained under constant medical observation. Inspection of this group has included routine screening for some of the most frequent health consequences of radiation exposure—such as cancer and developmental defects. Thorough observations failed to determine any negative impact from the accident on the health of those subjected to radiation. Some of the survivors showed evidence of radio-phobia, in other words the fear of radiation exposure and its potential for adverse health effects. Medical observation of the exposed group, including their offspring, continues to take place to this date.

Several independent studies estimated that the average radiation dose from the accident was less than 100 μREM, which is about a third of the natural background radiation. As such, human loss and severe health consequences were avoided due to a comparatively small amount of released radiation. Timely medical intervention and appropriate ecological measures further assisted in preventing adverse outcomes.

Chernobyl Nuclear Power Plant Catastrophe

Construction of Chernobyl' nuclear power plant (**CNPP**), located 130 km (81 mils) from Kiev (the capital of Ukraine), began in the 1960s. At that time in the history of the former Soviet Union, there was a strong push for a wide use of the so-called "peaceful atom", specifically in nuclear power engineering. The decision to build the CNPP, despite some timid objections from ecologists and other specialists, had to be unconditionally executed. Within a relatively small time frame, six nuclear reactors were manufactured. Regretfully, design deficiencies and imperfections of the fourth nuclear reactor did not influence the deadline for its operational start, which was firmly designated by various government establishments and communist party authorities.

Recently declassified archives revealed that a possibility of a nuclear reactor breakdown was suspected almost two months prior to the CNPP accident. This information was highly restricted and generated no actions.

The nuclear explosion occurred on the night of April 26, 1986, during a planned system test on the fourth reactor of the Chernobyl' nuclear power plant.

Its cause was a sharp surge in pressure and core temperature lasting only about 20 seconds. As a result of the subsequent explosion, the roof was torn away and the fourth reactor was completely destroyed, freely releasing radionuclide's into the atmosphere. Following declassification of official archives, additional facts leading to this catastrophe became known. All events contributing to this tragedy received a symbolic comparison. In the final analysis, technical shortcomings, compounded by a human factor, caused the most devastating nuclear plant catastrophe in the world. Later, a Nobel Prize laureate *Hans Bethe* said: "*Fundamentally faulty, having built in instability* ".

The accident took place Saturday night. During daytime, residents from a small town of Pripyat (location of CNPP) and surrounding areas enjoyed excellent spring weather by spending the day fishing, at picnics, etc. Children played in various gardens and parks. Life was moving forward at its usual speed on an off day.

A
The CNPP before (A)

B
and after (B) accident

Only a small group of specialists were called to work in connection with the accident. Power plant emergencies occurred in the past and, as a result, the most recent incident failed to cause significant alarm. The evacuation of people began only 36 hours following the accident, in the areas nearest to the nuclear reactor (radius of 30 kilometers-19 mils).

The first official information about the incident on CNPP appeared only on the third day following the explosion. A brief communication in an extremely optimistic tone was published in a state newspaper "Soviet Ukraine." Similarly scant information appeared in the national newspaper. The head of the government at that time, M. Gorbachev, was vacationing in the Crimean region (southern Ukraine) and was unaware of the incident (!?). In contrast, the President of the United States J. Carter arrived to the Three Mile Island immediately following the accident.

Subsequent information provided by national and state government establishments was characterized by unjustified and criminal optimism. These government sources assured people that the situation was under control and that there was no reason for panic. No concrete information about locations and types of radioactive contamination (soil, plant, water, etc) was provided to the public.

Even prior to the government's announcement of the CNPP explosion, employees from the physics institute at the academy of sciences in Ukraine noticed a sharp increase in background radiation, while conducting routine testing in the Kiev region. After conferring with senior personnel, this finding was attributed to the dosimeter malfunction.

On the 27th and 28th of April, 1986, various medical establishments in Kiev received an order to prepare for victims of the CNPP accident. However, no recommendations were made with regard to preparation specifics. During the same time, residents of Kiev and adjacent regions started to notice large numbers of buses, filled to capacity with people, moving into the city from the west. A portion of these buses were driven to the urban baths, where "cleansing" of victims was performed (people showered, yet were not given change of their clothing and footwear). One of such baths was located in a heavily populated location, in the center of the city opposite a national stadium. Contaminated buses were parked on the streets of Kiev for many hours. Following decontamination, people dispersed wherever they could (some stayed with relatives, others left Kiev, etc).

From the first day following the CNPP explosion, elevated background radiation levels were recorded in the **Scandinavian countries, Poland, Czech Republic, Austria, Southern Germany**, and **Northern Italy**. Following a change in regional wind patterns, *background radiation* increased **in the Balkans, Greece, Turkey,** and even **in two states of the U.S**. This was picked up and reported by numerous foreign radio stations (the "enemy voices"). Soviet citizens were categorically advised against listening to foreign reports. Nevertheless, information about the CNPP accident gradually spread and became well known to some portions of the population.

Despite providing criminal and outrageous lies, higher members of the Soviet party in Moscow, Ukraine, and Belorussia were well informed about the heavy nature of the accident, along with some methods of personal prophylaxis and the need to evacuate children younger than 14 years of age. This information was available under the "top secret" heading. People of the whole country, including the inhabitants of afflicted regions, were criminally reassured by persistent calls to resume normal daily routines, to participate in the May 1st demonstration (International Workers' Day), and to parade in celebration of the Victory Day (May 9th). To perceive this deceit and fabrication was very bitter and difficult.

So what really took place at the CNPP?

As a result of the explosion in the fourth nuclear reactor, 100 to 150 million curies of radionuclide's entered the atmosphere. This number is 300 times the amount released from nuclear blasts in Japanese cities of Hiroshima and Nagasaki in 1945. Radioactive particles were present in three physical forms—gas, aerosols, and solid fragments. Two separate reactor explosions contributed to an unprecedented scale of this accident. Initially, a combination of personnel error during the reactor testing, compounded by underlying design flaws, caused a significant power surge in the reactor core. This resulted in the first explosion, which lifted off the reactor cover plate and expelled fission products into the atmosphere. A subsequent explosion took place several minutes later, causing the reactor fuel products and graphite from the core to catch fire. The burning graphite contributed to a prolonged release (over 15 days) of more than 520 different radioactive nuclides, which contributed the bulk of the overall radiation fallout from the accident.

The most powerful release of radioactive products was observed in the first two to three days following the explosion at CNPP. On the sixth day, radiation levels started to increase due to heating inside the active zone to more than 2,000 degrees Celsius. On the ninth day the output of radiation products grew to 60% of the initial release. Following reactor cooling, levels of released radiation steadily decreased. However, two additional periods of radiation increases occurred between 8th-11th and 14th-17th of May, with the maximum peak on May 16th, 1986.

Since the reactor was no longer contained following both explosions, released radioactive materials rapidly ascended into the atmosphere, forming a radioactive cloud. Atmospheric winds subsequently spread this cloud in various directions, depending on the force and patterns of airflow. Subsequent radioactive fallout in different geographic regions occurred during periods of rainfall.

The most prevalent radioisotope during early stages after the accident was radioactive iodine (I^{131}). Subsequently, the radioisotope spectrum expanded to include cesium, ruthenium, strontium and many others. Of note, the radioisotope spectrum has changed and will continue to evolve over time. For instance, relatively recently an aggressive isotope Americium was discovered in a zone of the accident. This particle is characterized by a long half-life period as well as high degree of penetration. It is predicted that in approximately 50 years, Americium will occupy the most prominent place in the Chernobyl's radionuclide spectrum.

Large quantities of chemical substances were released along with radioisotopes into the atmosphere. Some of these (such as lead) were also used to extinguish fire in the fourth reactor. As it later turned out, the use of lead for this task was absolutely unfounded.

Radioactive materials were carried by wind into Ukraine, Belorussia and other regions of the four western Soviet Republics. Information also exists supporting that wind direction was altered toward the West—and away from Moscow—with the aid of special methods.

In *Ukraine*, 12 regions acquired a significant radioisotope pollution: **Chernigov, Charkas, Chernovits, Ivano-Frankovsk, Kiev, Kirovograd, Ravens, Sums, Ternopol, Vinnitsa, Volyn, and Zhitomir.** Populations **of Zhitomir, Kiev and Ravens regions** received the highest overall radiation dose.

70% of the total radiation fallout occurred in **Belorussia**, primarily affecting the following five regions: **Brest, Gomel, Grodno, Mogilev, and Vitebsk**.

In *Russia*, pollution from radiation was registered in regions of **Bryansk, Kaluga, Orel, and Tula**. Approximately 60% of radiation fallout affected the rural population.

According to the data from Ministry for Atomic Energy and the Nuclear Industry in the former Soviet Union, total area affected by radioactive materials exceeded 130 thousand square kilometers, with the population of 20 million people. Data from the Ministry of Public Health **of Ukraine** confirmed that more than 17 million people lived in the involved territories. In addition, more than 4.6 million hectares of fertile ground were struck.

In *Belorussia*, more than six thousand square kilometers of soil (half of which was fertile) was rendered unusable secondary to deposited radiation. The population of contaminated areas comprised more than two million people. Villages became deserted and kindergartens and schools were closed. However, only in 1990 the Supreme Council of the Republic declared affected territories as zones of ecological disaster.

In *Russia*, contaminated areas comprised more than 59 thousand square kilometers, including 2.9 million hectares of agricultural land and 1 million hectares of forestland. Almost 1.8 million people continue to inhabit these territories. In all, only 52 thousand people were relocated from contaminated areas. Unfortunately, people continue to live in places still polluted by radiation, despite almost 20 years after the accident. Experts estimate that almost 7 million people remain under conditions of radioactive contamination.

It is important to emphasize that sources of information on the impact of radiation fallout (such as the force of explosion, polluted areas, number of victims, etc) differ due to a number of subjective and objective variables. Subjective reasons include the incompetence, deceptiveness and falsification, as well as cover up of true facts by government organizations of the former Soviet Union and subordinate party organs in Ukrainian, Belorussian, and Russian Republics. The unauthenticated of information persists to this day in the currently independent states of Ukraine, Belorussia and Russia. Objective reasons include low quality and shortage of radiation monitors, insufficient level of training of technical and medical personnel, and an insufficient financial base. In addition, to a certain degree the variability of fallout data may be explained by differences in half-life spectra of radioactive particles in various locations. Natural as well as artificial displacement of earth's crust (large-scale constructions, environmental calamities, etc.) may have also contributed to these differences.

In terms of the health impact on people, four distinct phases of radiation fallout must be recognized and independently discussed:

1. *Immediate phase*: characterized by the development of a radioactive cloud. During this process, very high doses of radiation entered the organism mainly through respiratory organs. The bulk of this radiation affected the lungs, thyroid gland, and gastrointestinal tract.
2. *Iodine phase*: lasted 1-2 months, during which the thyroid gland absorbed most of the radiation.
3. *Two-three year phase*: primarily characterized by external gamma-radiation from short-lived radioactive particles (Ba^{140}, Cs^{141}, Cs^{144}, etc.). During this time frame, internal radiation occurred as a result of radioactive Cs^{134} and Cs^{137}, which entered the organism via food products and water.
4. *Long-term phase:* lasting from several years to tens and hundreds of years, characterized by radioactive action of cesium, plutonium, and strontium (**Table 1**).

Radiation fallout caused by the CNPP accident had a number of other peculiarities. They included:

- The patching (non-uniformity) of deposited contamination;
- Diverse composition of radioactive nuclides, in many respects dependent on the distance from the damaged reactor;
- Different mechanisms of radionuclide distribution from soil to plants and, subsequently, to people;

Many secret documents became available to the public 15 years after the accident. Only time will tell how fully they were revealed. The extent of the Chernobyl' catastrophe continues to be re-evaluated.

Following the accident, persistent absence of accurate information led to appearance of various rumors and incorrect recommendations. For instance, alcohol was recommended for prophylaxis and treatment of radiation-associated injuries.

Numerous decontamination projects in affected areas were improperly conducted. Houses were cleansed with water from fire pumps; however, this water was allowed to flow freely on the surface of streets, gardens, etc. Leaves were collected primarily by students and were subsequently burned in populated locations. A considerable number of other similarly disgraceful practices were performed in contaminated areas.

Isolated scientific and practical publications (the so-called methodology letters and information bulletins) were largely inaccessible to the population. Physicians had restricted access to this information and were required to obtain special permissions.

Soviet bureaucrats of all levels, with the participation of the KGB, hastily pointed out negligence by the officials at CNPP, without troubling themselves to look for all the causes of the accident.

Immediately following the accident on CNPP, work began to stop the leakage of radioactive materials. Deficient in specialized training and equipment, designated workers risked their lives and did everything possible and impossible to slow the consequences of the accident. The only available "robot" soon began to malfunction. Despite offers from other countries, Soviet government refused to accept foreign "robots." Soldiers without expert training and knowledge of radiation protection were sent to replace the broken "robot" and continued its work. Comedians with bitterness called them "bio robots".

Soon after the accident, a special shelter called sarcophagus was constructed around the reactor.

Shelter ("Sarcophagus") around damaged reactor

However, leakage of radioactive materials persisted. After several years it became obvious that the sarcophagus failed to fully contain the site and ensure safety. For the government of Ukraine, it became necessary to set aside their pride and ask for help from developed countries and influential international organizations (World Health Organization (WHO), United Nations (UN), and many others).

As far as quality and in good faith is exercised in future. To day will again discussed projects cover the fourth reactor. At the initiative of the Commission of the European Community and the Government of Ukraine developed and launched the project "Shelter" (Shelter Implementation Plan-SIP). In late 2000 under pressure and at great financial support from Ukrainian President Leonid Kuchma ordered the closure of the fourth reactor of the station. Those dates again sidled away. Over the years continuo work aimed at improving the sarcophagus. It was only in November 2008 was signed the instrument of his appointment. Shelter. Shelter fourth project involves the conversion of the existing sarcophagus into a safe and environmentally stable system, called confinement. Project life-time—100 years.

It is important to emphasize that even after the closure of CNPP, the danger of further pollution with radioactive materials will remain. This could occur via several means, such as flooding with the ensuing spread of contaminated water, or wild fires consuming forests affected by radiation fallout. These must become topics of immediate attention not only in the affected countries, but also in the entire world. Fortunately, these agendas are actively promoted by prominent scientists as well as influential international organizations (WHO, UN, etc.).

The statue of Prometheus, a tamer of fire, stood in the center of Pripyat.

After the accident was installed near the destroyed reactor as a reminder of catastrophe unbridled power.

In the next years in memory of the events and people of Chernobyl catastrophe built monuments and memorials.

Monument in memory of the Chernobyl victims (Rivne, 2006)

Also created films, works of art. In the city of Kiev opened a museum.

Accident on a Nuclear Power Plant in Japan

This accident took place on September 30th 1999 approximately 120 kilometers (80 mils) northwest of Tokyo.

Nuclear Power Plant in Japan

It occurred as a result of a mistake by two workers while loading radium into the reactor. Two other people were present in the immediate vicinity. Dissymmetry measurements showed that resultant radiation dose ranged from 24.5 to 39 GY. Additional 229 employees were present on premises of the power plant. Their radiation dose constituted 0.07-0.48 μSV. Another 350 people lived in a radius of 250 meters and were exposed to 0.01-0.21 μSV. All employees and inhabitants of the surrounding areas were quickly evacuated to safe locations. For the population living within 10 kilometers of the accident site, a recommendation was made to stay inside buildings. All schools and stores were promptly closed.

This accident quickly received response from international organizations and various societies. Despite the timely and intensive medical response, five people situated near the reactor died from acute radiation exposure and associated complications. Seven people sustained body burns.

Lessons from this accident were thoroughly analyzed, published, and appropriate changes were implemented.

Relatively minor and insignificant accidents on nuclear power plants continue to occur in Ukraine, Russia, U.S., etc.

Nuclear Weapons

Atomic Bomb

Following the discovery of nuclear fission and subsequent development of nuclear energy, several developed countries raised a question about the possibility of its application for military purposes. Prominent nuclear scientists in Germany, Russia, and the United States were involved to address this issue. Their work was carried out under strictly secret conditions.

In this period of time for Nazi regime of Germany was logical to try to use these discoveries for military. Physicists L. Szilard (1898-1964) and E Teller (1908-2003) is well understood. Forced to emigrate to the United States, they decided to discuss their concerns with A. Einstein. The great physicist understand what threatens humanity with the creation of the atomic weapons. it was decided to apply to the 32 U.S. President Franklin Delano Roosevelt. August 2, 1939 Einstein sent the President a letter, which clearly demonstrated the need for a nuclear bomb.

According to direction of the President August 13, 1942 established a special unit of the engineer troops in the desert.

The project was named "Manhattan" it was directed Brigadier-General L. Groves and scientific R. Oppenheimer.

July 1945 created the atomic bomb before the unseen forces proposal to deploy in the country of having nuclear weapons.

The history of the atomic bomb in the former Soviet Union refers to 1946. An outstanding scientist **I. Kurchatov** discussed with Joseph Stalin situation of so-called uranium project. In August 1949 was held testing bomb.

Several nuclear scientists feared that their discovery could lead to grave consequences for the humanity. One of the inventors of the atomic bomb.

Professor R.**Oppenheimer,** stated: *"All subsequent generations of people, while admiring the discovery of thermonuclear energy in the 20th century, apparently will never cease to curse the time, when creations of a human mind were subordinated to the aims of the invention of a terrible weapon of destruction."*

Later, **G. Wiener** wrote: *"Nuclear bombing—it is the fear of humanity for the scientific discoveries".*

Some politicians just as seriously perceived this potential issue. Thus, **W. Churchill** said: *"The Stone Age may return on the gleaming wings of science".*

Presently, the following countries officially possess nuclear weapons: U.S.A., Russia, England, France, China, India, Pakistan, and North Korea. This list continues to expand as a result of officially provided information as well as secret intelligence and spy data. Despite repeatedly announced bans on the arms race, it continues to take place.

Nuclear weapons may be manufactured in the form of bombs, projectiles, and torpedoes. Only recently, the United States designed an atomic bomb specifically for the destruction of bunkers. It may not be ruled out that in the near future applications of nuclear weapons will be further expanded.

So what is the atomic bomb, the tragic outcome of which is already known to humanity.

Depending on the filler—substances that condition and catalyze radioactive decay—three types of atomic bombs are distinguished:

* Uranium or plutonium;
* Helium and hydrogen;
* Thermonuclear;

The effect of atomic bombs foremost includes an enormous destructive force, unrivaled by anything in the world. Significant danger lies in accompanying high temperatures (more than 2,000-3,000 degrees Celsius) and ultraviolet radiation. Acute radiation effect comprises 5%, while residual radiation contributes 5-10%. A nuclear explosion results in the release of alpha, beta, gamma, and neutron rays and particles. Alpha and beta rays are adsorbed by air and do not reach the earth's surface.

The atomic explosion may injure a human organism by causing direct trauma, burns, or radiation exposure. When occurring in combination, heavier consequences ensue for the victims. However, fatal outcomes may arise as a result of only a single injury.

The atomic bomb was used for the first time by the United States at the end of the Second World War. On August 1st, 1945, the Japanese city of Nagasaki (population 195,000) and then five days later the city of Hiroshima (population of 255,000) were bombed. As a result of nuclear bombings numerous buildings were destroyed. Mass casualties totaled 39,000 in Nagasaki and 66,000 in Hiroshima. Tens of thousands of people suffered heavy trauma, burns and radiation injuries. Among victims were citizens of other countries, including the former Soviet Union. Of note, casualty numbers vary slightly depending on the source of information.

A   B

Atomic bombs (A—"Fat men", B—"Baby")

Radiation cloud generated by exploded bombs resembled a giant mushroom.

Cloud like mushrooms by explored bomb

It was established that the overwhelming majority of deaths were caused by the initial shock wave. The death toll stemming directly from immediate radiation exposure.

Unfortunately, another tragedy of this magnitude cannot be completely excluded for the future. One can only believe in the wisdom and prudence of governments of those countries in possession of nuclear weapons. Of note, a possibility of accidental nuclear war outbreak also exists. This tragic disaster may arise from several catastrophic causes (errors in systems processing military information and controlling combat control, technical failures and malfunction in combat systems, erroneous or unauthorized action.

In memory about the terrible tragedy, built a lot of memorials, which lists the names of the victims. On anniversaries of the monuments and memorials produce paper cranes, the number of which corresponds to the number of the victims. On much of them is touching :" Sleep in peace, this does not happen again."

Not so far from the epicenter of atomic bomb opened International Parks and Museums.

Monument in memory about the tragedy

In 1990, the authoritative international organizations agreed on non-proliferation of nuclear weapons and ending the tests. Everybody like to believe in the wisdom of governments of countries with nuclear weapons.

One can not mention the possibility of so-called accidental outbreak of nuclear war. Such a tragic accident could be caused by several reasons—the error information processing systems and command and control.

Testing of the Hydrogen Bomb (H-bomb)

A 17-megaton H-bomb was tested in 1954 on Marshal Islands (USA). The explosion of this bomb occurred primarily underground and extended across 100 miles (160 km). Direct influence of radiation, blast wave, and high temperature were not observed. Nevertheless, an increase in natural background radiation was registered in the area. In response, long-term monitoring of this location was immediately commenced.

Hydrogen Bomb

Due to potential release of radioactive materials that accumulated in the earth as a result of the explosion, local population was permanently evacuated into safe areas. Consequences of radiation exposure were not detected in resettled individuals.

However, during subsequent years, increased incidence of hypothyroidism and thyroid cancer was detected in children. During a 10-year follow-up, no cases of leukemia were discovered in this cohort. Ongoing observation of this population continues to take place.

The decision to establish super bomb in the Soviet Union, Nikita Khrushchev accepted to show the imperialists that we are able. Dimensions super bomb impressed. The explosion occurred on October 30 1961. at 11:32 Moscow time. The flash was so bright that it can be observed from a distance of 1000 km (625 mils). In its place arose an orange ball of glowing gas that absorbs tens of kilometers of space. Giant "mushroom" rose to an altitude of 65 kilometers.

Eyewitnesses described the explosion as the brightest flash in the 300-kilometer distance. Later, they heard the distant and powerful roar. The light flashes coming from a huge fireball. Despite considerable height 4 km

(2.5 mils), reaching the ground, a ball of fire continued to grow to the size of about 10 km (6.2 mils) in diameter.

With capacity of 50 Mega-tone area of total destruction was a circle of 25 kilometers (15 mils).

The flash was so bright that it could get third-degree burns to necrosis of the upper layers of the skin on distains 60 km (39 mils).

After the explosion due to ionization of the atmosphere for 40 minutes was interrupted radio news.

Secret tests of hydrogen and atomic bombs have also taken place in other countries.

"Dirty Bomb"

Terrorism represents a complicated and multifaceted concept. Nevertheless, its tragic consequences are well known to the world. Predictably, terrorist organizations would seize an opportunity to use radiation for their fatal aims. This would become a frightening weapon in the hands of terrorism.

Several leading American scientists (Teodor Teylor, Eugen Eyster, William Maraman, and Jacob Wechsler) concluded that use of nuclear weapons for terrorist purposes represents a feasible concept. However, not everyone is willing to agree with this conclusion. This stance, in turn, is negligent and potentially very dangerous.

The so-called "dirty bomb" represents the most practical application of nuclear terrorism.

According to some estimates, the probability of radiation the next 10 years is 19%, a use of "dirty bomb"—40%.

Manual scattering of radioactive materials forms a less reliable option.

According to estimations, a small nuclear device detonated at daytime in New York could lead to casualties on the order of 500,000 people.

So what is a "dirty bomb?" It could be considered a version of the nuclear bomb, containing a combination of a conventional explosive (such as dynamite) and a radioactive substance. This bomb functions by explosive-induced airborne radiation and contamination by a radioactive material. The latter could be in the form of cobalt[192], strontium[90], plutonium[238], americium[241], etc. Sizes of "dirty bombs" vary from large to small, allowing easy placement into a bag, briefcase, etc.

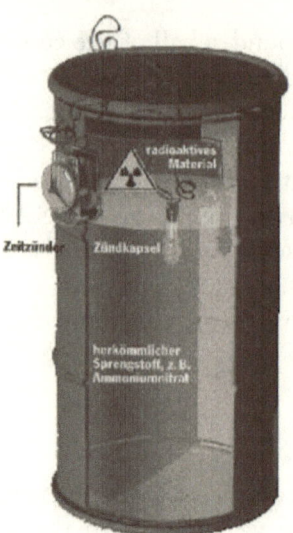

"Dirty bomb"

Manufacturing a "dirty bomb" does not pose extremely difficult obstacles, since terrorists may obtain principal components—including radioactive materials—from a variety of sources (laboratories, nuclear power plants, radioactive wastes, etc.).

Enriched uranium can be found in more than 130 establishments in 40 countries worldwide. Unfortunately, security systems used for their protection are not always the state of the art. Cases of stolen uranium have occurred repeatedly. The probability of theft highly increases during the times of chaos, as exemplified by what took place in Iraq.

The so-called hot particles—sufficiently small radioactive elements that may be dispersed by the wind—present a particular hazard from a "dirty bomb" explosion. They may not necessarily affect the background radiation. However, upon entry into a human organism, such particles become forever entrapped. Cells of various organs, now in the immediate vicinity of a radiation source, may receive heavy doses of radioactive exposure with all the ensuing negative consequences.

Regretfully, the method of "dirty bomb" assembly is known to large terrorist groups. According to the data from the U.N., a "dirty bomb" was tested in Iraq in 1987. Fortunately, the intensity of radiation was sufficiently low to generate any damage.

In contrast to nuclear bombs, the primary objective of "dirty bombs" lies in causing the spread of radioactive material. Significant hazard arises from radioactive particles further dispersing by various means (automobiles tires, water, etc.) across large areas. The explosive force of such bombs is otherwise insignificant.

"Dirty bombs" may be used to destroy smaller targets (houses, subways, etc.). They may also be utilized to contaminate water reservoirs, various food sources, etc. Possible consequences of "dirty bombs" depend on numerous factors, such as the quantity and quality of radioactive materials as well as environmental conditions, which ultimately determine areas of dissemination. Distance from the point of impact represents another significant consideration, since it is inversely proportional to induced damage. These complex issues are subject to further thorough studies.

Radioactive particles may enter a human organism during the respiratory cycle (inhalation mode), through the gastrointestinal tract, and across injured skin. External and internal radiation exposure may occur as a result of a "dirty bomb" explosion. Acute radiation injury remains a distinct possibility. Significant hazard arises from radioactive contamination of the environment, especially food. Therefore, decontamination of affected territories carries utmost importance, in order to avert ensuing effects of radiation. Some believe that chronic radiation generates worse consequences on the organism than acute radiation encountered during the explosion.

Despite all of the above, some Russian, Ukrainian and Belorussian scientists assert that the effect of a "dirty bomb" may be equated to the creation of panic among people (the "weapon of mass panic"). Disappointedly, this unsubstantiated opinion finds support in some government institutions and is negligently propagated by the media.

A number of governments, including the U.S., focus on undertaking specific measures against nuclear terrorism. Not long ago the United Nations affirmed a resolution prohibiting the use of radiation for terrorist purposes. There is some information about the preparations for the use of "dirty bombs" (London, 2002). In 2005, the U.S. was detained terrorists prepared to blast a "dirty bomb." These facts, the information is not limited. Terrorists reported timing of the "dirty bombs." In order to terrorism can be used and other isotopes.

In November 2006 was poisoned with polonium radioisotope fled abroad former worker of KGB Alexander Litvinenko. He had received dose was 42 Gr. The victim died 22 days later.

Chelyabinsk Nuclear Catastrophe

In 1945, in the former Soviet Union, a nuclear complex called "Mayak" was constructed in a small town of Kishtim (Ural Mountains, approximately 60 miles from the city of Chelyabinsk). The primary objective of this enterprise was to obtain enriched uranium and plutonium for military purposes. Only high-ranking government and military officials knew about the existence of this complex

Chelyabinsk Nuclear enterprise "Mayak"

In the 45 years that followed, the growing town of Kishtim was closed to outsiders. This meant that entry by non-workers and foreigners was strictly prohibited. Employed personnel as well as residents of this town were under constant observation by the committee of national security (KGB). From 1948 to 1992, approximately 73 tons of plutonium were produced.

Small accidents periodically involved the nuclear reactor from the beginning of its operation. However, this information was never officially publicized.

On the 26th of September, 1957, a large mishap occurred on the nuclear power plant. The cause was a combination of insufficient personnel training and a deficiency in reactor design. Even prior to the accident, nuclear waste was discarded into the Ticha River and nearby Karachai Lake. As a result of the plant accident, the bulk of radioactive plutonium ended up in the Ticha River and Karachai Lake, which soon turned into a swamp, and further promoted the escape of radiation into the atmosphere.

In all, 108,000 CI of radioactive materials was released. Strontium[90], which has a long half-life, comprised the majority of released radio nuclides. Radioactive substances were present in three physical forms—solid, gas, and liquid. The contaminated territory was fenced with a barbed wire.

Evacuation of people from the most contaminated areas began only after 7-10 days. The whole event was kept highly classified and no clear information about the accident itself or measures of protection and prophylaxis against radiation were provided. Even the results of victims' blood and bone marrow analyses were strictly classified until 1992.

As a result of this accident, approximately 124,000 people received high radiation doses, ranging from 0.35 µSV to 17 µSV. Furthermore, significant increases in background radiation were noted in nearby regions of Chelyabinsk and Tumen. Radioactivity of water samples from the Ticha River was 400,000 BQ (cesium^{137}) while radiation content of soil samples was 120,000 BQ (strontium^{90}).

Adolescents participated in the cleanup of the accident and were referred to as "young liquidators." To this date, decontamination of affected territories remains to be completed. Some firmly believe that this location represents the most contaminated area in the world.

In the 40 years of reactor's existence, 10,000 people received high radiation doses, with 4,000 suffering various degrees of radiation sickness. As a result of the last—and most significant—accident on "Mayak," 42 people experienced acute radiation sickness. In contrast, chronic radiation sickness was discovered in approximately 1380 people. Almost 200 people sustained radiation burns. In subsequent years, 123 victims developed plutonium pneumosclerosis.

Initially, the disclosed information about the scale of radiation fallout and its impact from this nuclear accident was difficult to believe. Released numbers appeared excessive, until a documentary about victims' lives came out in 2005.

For almost 35 years—or until 1992—specifics about the accident remained classified. Until that time, practically nothing was accomplished toward the treatment and prophylaxis of those affected by radiation injuries. Even currently conducted measures remain absolutely insufficient. The region's environment remains contaminated while people continue to live in affected areas in very poor social conditions. Government-proposed relocation of people represents an unrealistic notion. Routine observation and follow-up of victims as well as medical and social rehabilitation issues remain poorly addressed.

In 1990, a number of influential international organizations passed a resolution about non-proliferation of nuclear arms and cessation of nuclear weapon tests. And only in April of 2005, local authorities decided to shut down "Mayak." Regretfully, this failed to resolve fallout-related complications, which accumulated during the years marked by lies and inactivity.

Only in April 2005, local authorities decided to close the facility "Mayak". Unfortunately, this does not solve the problem of security and all the complexities that have accumulated over many years of lies and inaction.

The function of "Mayak" should be unmistakably qualified as the use of nuclear energy for military purposes. This fact has become obvious for a large circle of specialists as well as the population.

Radioactive Waste

Human health hazard from radioactive wastes continues to increase fairly rapidly in the entire world, including in our country. This is caused not only by a steady growth of radiation use in various spheres of life, but also by difficulties associated with radiation cleanup, transport and burial.

Similar to original sources of radiation, radioactive wastes exist in three physical forms—solid, liquid, and gas.

Container for radioactive waste

The majority of radioactive waste liquidation programs are either poorly conducted or not conducted at all. All branches of science and engineering that utilize radioisotopes as well as nuclear power plants serve as a source of radioactive waste, with different associated levels of radioactivity. Nuclear weapon tests and nuclear submarines also have a prominent role in generating radioactive waste.

High radiation-content nuclear wastes are especially formed during processing of used fuel from nuclear reactors. Radioactive waste may enter into a human organism by several means, such as via the inhalation of aerosolized particles and water vapors, external radiation, and food (through the gastrointestinal tract).

As already noted, decontamination, transportation, and reliable protection of radioactive wastes presents significant challenges. Different methods of radioactive waste processing facilitate their transport: consolidation, verification, calcinations, cementing, etc.

Microbes termed kineococcus radio tolerance were recently used in the United States to neutralize radioactive wastes. Similar to other scientific breakthroughs, properties of these particular microbes were discovered accidentally. Their wide use in the future remains very promising.

Radioactive waste is disposed using specifically designed containers in assigned locations, preventing any future extraction or leak into the environment. Zones of radioactive waste disposal are systematically monitored in the United States. This includes routine ecological maintenance (measurements of radiation levels, decontamination, etc.). Enormous danger of radioactive wastes lies not only in their detrimental effect on people, but also in their potential use for terrorist purposes.

In zones containing radioactive waste, an ongoing evaluation of population's health and health-improvement measures receive significant attention. One such area is located in the state of Texas approximately 17 miles north of Amarillo, where nuclear weapon tests were conducted in the 1950's. Different specialists, including radio-physicists, have thoroughly evaluated this region in subsequent years. Assessment of the health impact has occupied a central place in the follow-up of this area's population.

Another zone with increased levels of radioactivity exists approximately 33 miles from Columbus (Ohio), around the location of a nuclear power plant. Together with various ecological initiatives, numerous programs simultaneously evaluate the health of a local popul1ation.

Similarly, an area exists approximately 16 miles northwest of Denver (state Colorado) where studies with radioactive materials were conducted in the past. This territory was decontaminated by 1992. Since then, routine medical observation of those living in this region has taken place.

Some cases are known in which absurd circumstances caused radioactive exposure of people. In one such incident, scavengers dismantled a metal canister from a radiotherapy machine at an abandoned Brazilian cancer clinic and left it in a junkyard. During this process, a 20-gram capsule containing radioactive cesium was ruptured. As a result, 14 people received high radiation doses and 4 people subsequently died. 229 additional people received lower radiation exposures. 85 houses had to be demolished during decontamination and victims of this incident received large compensations.

Similar cases have occurred in other places, including in our country. Therefore, reliable protection of radioactive wastes remains of utmost importance. Unfortunately, this issue is yet to be entirely resolved.

For storage, transportation and disposal of radioactive waste in Ukraine, Belarus and Russia created the set of instructions and orders, most of which are not executed. Today the problem of transportation and care radioactive waste is far from being solved.

DESTRUCTIVE MECHANISMS OF RADIATION

Research evaluating radiation effects in living organisms has been conducted for over a century. As a result, a number of mechanisms underlying radiation-induced injuries have been elucidated. Long-lasting effects of radiation in people are principally documented from follow-up of radiation exposure victims and of nuclear plant personnel handling radioactive materials. Unfortunately, sources of radiation exposure continue to increase. This stems from ongoing development of novel technologies, which add to the myriad of existing radioactive products utilized in various branches of science and industry.

Progress and achievement in many different areas of science and technology (physics, chemistry, physiology, pathology, etc.) has significantly improved the capabilities of radiation research. Already today, available research data allows to formulate and better understand specific mechanisms by which radiation induces injury in people.

Damaging effects of radiation principally depend on three factors—duration of action, distance from the source, and containment of the source.

Three factors damaging effects on radiation

The degree of injury from exposure to external radiation is directly proportional to duration, while inversely proportional to the distance from the source and thickness of protective layers.

In other words, longer distance from the epicenter of radiation as well as shorter duration of exposure cause considerably less damage. Moreover, adequate and reliable containment of radiation leak leads to smaller releases of radioisotopes into the environment, reducing ultimate levels of exposure.

When X-rays and gamma rays comprise a significant proportion in a spectrum of emitted radiation, immediate relocation represents an especially important consideration. Every doubling of the distance from the radiation source reduces the radiation effect four times. Since gamma rays disperse across much greater distances than other forms of radiation, such as alpha—and beta—rays, moving away from the source of radiation significantly reduces exposure. Similarly, better containment of radiation leak leads to fewer detrimental effects, especially in the presence of beta—and gamma rays.

In the early years following nuclear bombings, distance from explosion epicenters served as the main criterion in assessing the degree of radiation damage. It was firmly believed that larger distances firmly correlated with lower levels of induced radiation injury to people. Analogous estimations were also applied in cases of power plant accidents. However, in all these instances, variable geographic distribution of radioactive substances was not taken into account. At present, the pattern of radioactive dispersion is compared to the leaps of a kangaroo.

Over time, more reliable methods assessing quantitative and qualitative characteristics of radiation have been developed in order to accurately estimate sustained radiation injury.

Absorbed radiation dose represents the fundamental quantitative measurement of radiation effect on any object, including the human organism. It equals to the difference in energy, lost as a result of entry and exit of radiation through a particular medium. However, its direct determination is extremely complicated. Therefore, for practical purposes, indirect methods are better suited for determining the absorbed dose. They include emission particle counters, ionization chambers, and different types of dosimeters. Retrospective estimation of exposure dose remains an area of active investigation.

Detrimental effects of radiation on human health largely depend on means of radioisotope exposure of the organism.

Two main categories of radiation exposure are distinguished: *external* and *internal*.

External exposure may occur from elevated radiation levels in the environment (air, soil, reservoirs, and other) as well as from contact with contaminated possessions such as clothing, footwear, etc. Radiation may enter the organism through the respiratory tract during inhalation of aerosolized radionuclide's or the digestive system via contaminated food and water. Internal exposure through open skin wounds represents an especially dangerous form of radiation exposure.

When compared to external radiation, internal radiation caries more severe consequences. This is secondary to selective and strong binding of radioisotopes within cells of various tissues (**Table 5**). This functions to prolong their damaging effects. Internalized radioisotopes fail to extrude immediately and gradually decay, producing a sustained destructive effect inside an organism. Internal impact of alpha—and beta radiation may be especially significant. While this type of radiation energy travels only a few millimeters in tissues, it carries high ionizing potential. This produces a significant regional cellular injury. However, this type of radiation is practically ineffective against intact skin, which forms an effective barrier and prevents its penetration.

The establishment of correlation between various levels of radioactive exposure and their ensuing effects represents a significant accomplishment of nuclear biology and medicine. It is largely based on the experimental data as well as on ongoing surveillance of radiation victims.

A dose of radiation equivalent to 10,000 RAD (100 GY) causes death within several hours or days as a result of irreversible damage sustained by the central nervous system and other life-sustaining organs. Radiation exposure to 5,000-10,000 RAD (50-100 GY) leads to death in one to two weeks principally from internal hemorrhage in the gastrointestinal tract. 50 % of people exposed to 300-500 RAD (3-5 GY) die in a course of 1 to 2 months, principally from lethal effects exerted on all parts of the bone marrow. Primary radiation sickness develops from exposure to 150-200 RAD (1.5-2.0 GY).

Temporary infertility occurs in people exposed to 100 RAD (1 GY). A dose less than 100 RAD (1 GY) causes the *bone marrow syndrome*. It is characterized by bone marrow suppression as well as dysfunction of the spleen and lymph nodes. This may secondarily lead to internal hemorrhage, weakness, bacterial infection, and fever. Similar radiation exposure levels may also produce the *gastrointestinal syndrome*, which has the following manifestations: nausea, diarrhea, vomiting, dehydration, electrolyte disturbance, bleeding ulcers, as well as previously mentioned signs of bone marrow suppression.

Radiation dose of less than 5,000 RAD (50 GY) is characterized by the *central nervous system syndrome*. Typical symptoms include disequilibrium, altered level of consciousness, spasms, and a comatose state in the most severe circumstances. Usually at this dose, a concurrent shutdown of hematopoietic and gastrointestinal organs takes place, with accompanying characteristic symptoms as discussed above.

Acute external radiation exposure at levels ranging from 200 to 300 RAD (from 2 to 3 GY) causes hair loss and skin redness, with the latter typically resembling a sunburn. Radiation dose of 125-200 RAD (1.2-2.0 GY) may induce a disturbance of the menstrual cycle for various periods of time, and is observed in half of irradiated women. Exposure dose of 60 RAD (0.6 GY) leads to permanent infertility. Following exposures to 50 RAD (0.5 GY), benign tumors may arise in the thyroid gland. Radiation dose equivalent to 25 RAD (0.25 GY) could also be dangerous, especially in people with pre-existing medical problems (serious concurrent illnesses, diseases of the liver and kidneys, etc.). Genetic mutations—which increase the potential for congenital defects in the offspring—may arise with radiation levels as low as 10 RAD (0.01 GY).

Extent of organ-specific radiation injury heavily depends on the radio sensitivity of various tissues. For instance, temporary disturbances in the bone marrow, reproductive organs, and the gastrointestinal system appear with exposure levels 15-20 times lower than those capable of causing skin injury.

Data presented above is increasingly used for diagnosis and prognosis of radiation exposure, therapeutic and prophylactic measures, and subsequent long-term observation of victims and their offspring. Research investigating the role of absorbed radioactive materials in specific organs (for example, iodine in the thyroid gland) represents a relatively new focus in nuclear biology and medicine.

As previously discussed, radioisotopes may enter the organism through water, food, inhaled air, and much less commonly through skin or mucous membranes. A combination of these entry pathways may occur in the same individual (for example, concurrently through respiratory and digestive tracts).

Radiation may affect various biological levels (**Table 4**).

Level of Biologic Impact	Principal Radiation Effects
Molecules	Damage of enzymes and other systems, leading to the disturbance of biologic mechanisms
Sub-molecules	Destruction of cellular membranes, nuclei, and chromosomes
Cells	Suppression of cell division, death, transformation into cancer
Organs	Brain, bone marrow, intestines
Organism	Death
Population	Genetic code change

Table 4. Different biological impact of radiation.

Two broad categories are frequently used to grade health effects of radiation exposure in people: *stochastic and non-stochastic*. Stochastic effects arise as a consequence of prolonged exposure to low levels of radiation. In this setting, radiation disrupts numerous processes within the organism (see below), potentially resulting in development of genetic mutations and cancer. Non-stochastic effects represent a byproduct of short-term radiation exposure, most notably leading to acute radiation sickness.

It is important to recognize that signs of radiation injury may appear in various periods following the exposure. Moreover, manifestations of radiation injury may vary from one individual to another.

Radiation syndrome may exhibit the following symptoms and signs: weakness, weight loss, skin changes, inflammation of the external heart lining (pericarditis) and disturbance in the central nervous system, kidneys, liver and the gastrointestinal tract. Furthermore, inflammation of the lungs, sexual dysfunction, loss of vision, and physical retardation in children may also frequently be present.

Of note, radio sensitivity in children is considerably greater than in adults. In addition, the level of radio sensitivity rises with concurrent intake of antibiotics or chemotherapy. Radio sensitivity is especially high in the embryo and the fetus.

Concurrent consumption of alcohol and smoking may augment the overall impact of radiation. Furthermore, health consequences of radiation are also influenced by individual radio sensitivity or radio resistance. This factor determines the balance between the radiation-induced injury and the rate of recovery by the organism. Such processes may be studied on many different levels (**Table 4**).

Presented information testifies to the multifaceted nature of mechanisms and consequences of radiation exposure. The biological effect of radiation begins with the process called ionization, which changes the charge of atoms within cells. Carbon, oxygen, hydrogen and sulfur represent basic chemical elements of any organism. Oxygen caries the principal role in reactions generating energy from carbohydrates and fats. This energy is subsequently used by cells to produce enzymes, which function as catalysts (or accelerators) in the myriad of life-sustaining biochemical reactions. Ionization of oxygen may directly lead to a chemical breakdown of enzymes and other compounds found in tissues.

Radiation also targets fats and proteins, which are necessary for normal function of the organism. As a consequence, this causes molecular changes called mutations to appear. In turn, these predispose to the development of malignant tumors and may potentially be transferred to the offspring.

Mechanisms of radiation-induced injury in human organisms vary widely and depend on a number of external and internal factors.

Radiation transfers energy directly to tissues inside the organism. This energy uptake creates injury within nuclei of different cells, causing malfunction of their activity and, ultimately, death. The base material inside nuclei of cells is called chromatin, the main component of which is deoxyribonucleic acid (DNA). It carries a cellular genetic code, and injury at this level leads to mutations and change in the genetic (hereditary) apparatus. A genetic code represents the "program" for each cell. It caries a critical function of regulating cellular division and activity. A change in the number and structure of chromosomes represents a marker of genetic disturbance.

Under normal conditions, a critical system consisting of reparative enzymes functions to maintain and continually restore the integrity of genetic material contained within nuclei of cells. Radiation causes malfunction of this system, which firstly manifests in organs normally exhibiting rapid turnover of cells (bone marrow, reproductive organs, small intestine, and the thymus gland). The magnitude of injury not only depends on absorbed radiation dose, but also on the overall health status of irradiated organism.

It is well known that energy absorbed by biological tissues distributes unevenly. As a result, varying quantities of radiation energy transfer to different populations of cells within tissues. This uneven absorption pattern explains a number of special features found in radiation injuries.

Barriers against ionizing radiation do not exist in human organisms. As a result, radiation causes detrimental effects on function of all biological systems. Furthermore, ionized oxygen plays the key role in destruction of cellular enzymes—compounds that serve as catalysts for biochemical reactions.

The role of free radicals in causing the radiation injury has also been intensively investigated. Increased free radical content has been shown in blood as well as various organs of radiation victims. This strongly suggests their involvement in the development of radiation injuries. The quantity of free radicals increases proportionately with the degree of radiation damage sustained by molecules. These reactive elements destroy cellular components, ultimately causing injury to the whole organism.

Cellular radiation-induced injuries activate a number of intracellular protective mechanisms, which are designed to repair damaged molecules and adapt to a new cellular environment. However, capabilities of these mechanisms are limited and may become overwhelmed by sufficiently large exposure levels. In this setting, irreversible changes ensue, ultimately causing cellular demise.

Different sensitivity of cells to radiation in various organs may have practical significance. In general, organs with rapid cellular turnover are more sensitive to radiation exposure (for example, the hematopoietic system). Furthermore, different radioactive materials preferentially exert their influence on specific cells and tissues within the organism (Table 5).

Radionuclide	Most Sensitive Organs
Iodine[131]	Thyroid gland
Ruthenium[103]	Skin, lungs
Ruthenium[106]	Lungs, lymphatic system
Strontium[90]	Bones, bone marrow
Plutonium[239]	Lungs, bones
Cesium[144]	Lungs, lymphatic system

Table 5. Sensitivity of various organs to different types of radioactive nuclides.

Detrimental effects from prolonged exposure to low levels of radiation have not been researched as extensively. Chernobyl's catastrophe provides a significant reason to undertake such studies.

In theory, protracted exposure to small doses of radiation creates a more favorable environment for successful activation of compensatory cellular mechanisms. On the other hand, recurrent damage by different radioisotopes promotes propagation of injury in different organs. Harmful effects of radioisotopes may last for many years. Unfortunately, such radiation exposure likely affects the health of millions of people. Chernobyl's nuclear power plant accident created such conditions, secondarily allowing for an intensive study of this issue. Prominent scientists and various nuclear agencies have studied this area for the past more then 20 years.

Presently known mechanisms of radiation injury explain its multifaceted impact on people. They are also used for practical applications. Further investigation undoubtedly will improve treatment and prophylaxis of radiation injuries in people. Intensive research studying mechanisms causing radiation damage continues to take place.

CONSEQUENCES OF RADIATION EXPOSURE

In the 1960-70s, radiobiologists and radiation oncologists began to evaluate the long-term health effects from exposure to radiation. Radiation-induced long-term effect—diseases called radiation-induced (stimulated).

This led to improved awareness and better understanding of radiation-induced diseases, sharply increasing in the total number of recognized radiation victims.

Today, more than 20 delayed illnesses and functional disturbances are associated with antecedent radiation exposure. They include genetic mutations and their transmission to the offspring, development of cancer, immune suppression and deficiency, increased susceptibility to bacterial and viral infections, malfunction of the endocrine system, cataract development, temporary or permanent sterility, decrease in the average expected lifespan, delay in mental development, and other (discussed further below). Unfortunately, the list of radiation-induced illnesses continues to expand.

The material presented before demonstrates that several characteristics determine the extent of the impact from radiation exposure on human health. They include various mechanisms of entry into the organism, nature of the radiation source, and individual biologic differences among people subjected to radiation.

Radiation injuries and diseases may be divided into categories-
early and *late.*

Early Radiation Injuries

Symptoms of radiation exposure may arise from injuries sustained from various radiologic diagnostic and treatment procedures, especially during radiation therapy. Significant exposure levels to any radioactive source may cause acute radiation sickness, which manifests a typical spectrum of accompanying symptoms (reviewed below).

Radiation injuries associated with diagnostic and treatment procedures

As previously discussed, radiation research has contributed to a wide proliferation of radiation-based technologies in all branches of medicine. In the last decade, dramatic improvements have been made in diagnostic as well as therapeutic medical equipment. Safe application of these technologies is predicated on periodic calculation of emitted radiation, which allows for safety monitoring while minimizing injury to patients during various diagnostic and treatment procedures.

Utilization of radiation in medicine carries unlimited potential, with some exceptions. For instance, exposure to radiation remains contraindicated in a setting of early pregnancy and in those with increased sensitivity to specific radioisotopes (such as iodine). In addition, unindicted radiologic procedures should be categorically denied to insisting patients, who may deem them necessary from their point of view. While rare, any radiologic studies could potentially result in adverse outcomes, such as allergic reactions and burns to skin and mucous membranes

It is well known that radiation therapy for various forms of cancer represents the most effective available treatment option and, therefore, remains widely utilized. This applies to a number of cancers occurring in both men and women of any age. Of clinical significance, potentially serious complications may accompany and complicate various types of radiation treatment.

Individual radioactive load from exposure to radiation may vary. As a rule, most patients develop a range of symptoms of acute radiation sickness. These commonly include functional suppression of the bone marrow and other systems (pulmonary, cardiovascular, and gastrointestinal). Bone marrow suppression may lead to abnormalities in blood clotting (i.e. bleeding), anemia, and also infectious bacterial and fungal diseases of different organs (i.e. immune suppression). Such patients usually manifest skin changes (edema, reddening) and may develop hair loss.

Radiation side-effects may be ameliorated with special treatments, yet carry a lifetime risk of potentially recurring. Repeat episodes are particularly dangerous and difficult to cure. The process of recovery after radiotherapy can last up to two years. In addition, the use of very large radiation doses may lead to more serious complications. These include radiation injuries to the brain, lungs, and heart, as well as persistent nausea and vomiting.

Acute radiation injury of the brain (acute encephalopathy) usually manifests with nausea, headaches, and convulsions. These symptoms may appear not only at the very beginning of treatment, but also in subsequent weeks and months. Not surprisingly, this complication is more frequently encountered during radiation treatment for brain cancer. Similar to radiation injury of the bone marrow (myelopathy), acute encephalopathy may appear months and years after the end of treatment. Radiation injury of the heart is characterized by varying degrees of destruction sustained by the muscular wall (myocardium) and the pericardial sac (pericardium). Significant functional impairment may lead to poor cardiovascular prognosis. In general, frequency of complications increases in patients undergoing radiation treatment in conjunction with chemotherapy. Adverse side effects may additionally include radiation nephritis, prostatitis, cystitis, and proctitis.

All arising complications of radiation therapy demand immediate treatment response. In majority of cases, outcome remains favorable with a timely use of contemporary methods of prophylaxis and treatment. Death may occur in severe cases of radiation exposure as well as from delays in timely treatment of complications.

The resulting complications are subject to immediate treatment.

Acute radiation sickness (syndrome)—ARS

As previously discussed, the overall dose and entry pathway into the organism principally determine the magnitude of effects from radiation exposure. Protracted high radiation doses instantly result in death. Acute radiation sickness (ARS) generally occurs from a combination of several factors, such as the severity of the radiation dose -**0.7 GY**, injury specifically induced by penetrating beta and gamma rays, and a relatively short exposure time. Radiation levels capable of inducing ARS may arise from a wide variety of sources, including industrial accidents (cyclotrons, nuclear power plants, radioactive wastes), nuclear bombs and, possibly, "dirty bombs".

Widespread application of radiation energy led to a notable increase (hundreds of cases) in ARS. Before 1940s, only dozens of ARS cases occurred, usually in a setting of accidents occurring in scientific laboratories. **Table 6** depicts the relationship between the radiation dose and timing of appearance of various signs of acute radiation injury.

Dose (GY)	Signs	Time of Appearance
5-10	Changes in blood	Hours
50	Nausea	Hours
55	Weakness	Hours
70	Vomiting	2-3 weeks
75	Hair loss	2-3 weeks
90	Diarrhea	2-3 weeks
100	Bleeding	2-3 weeks
400	Death	2 months
1,000	Disturbances of gastrointestinal organs, internal hemorrhage, death	1-2 weeks
2,000	Injury to the central nervous system, death	Minutes, hours, days

Table 6. Radiation dose and timing of appearance of principal signs of acute radiation sickness (ARS).

Clearly demonstrates that severity of complications and fatal outcomes of ARS are directly proportional to the radiation dose. Researchers observing victims of the nuclear power plant accident in Chernobyl reached similar conclusions.

Extremely elevated radiation levels rapidly induce death, as has been commonly observed in a setting of nuclear bombing and atomic power plant accidents. As a result of Chernobyl nuclear power plant accident, 31 personnel members developed a very severe form of ARS. They all perished within several weeks after the accident. 140 survivors of acute radiation sickness received treatment in specialized hospitals of Moscow and Kiev. Their radiation dose generally ranged from 350 to 500 µSV. Acute radiation sickness has not been observed in the general population after the nuclear power plant accident.

Following nuclear bombings in Japan and radioactive accidents on nuclear power plants, victims of radiation exposure have been generally observed in dissimilar conditions. As a result, descriptions of ARS may differ to a certain degree. Nevertheless, principal signs appear to coincide and correspond well to results derived in experimental studies. Depending on the gravity of symptoms, four different categories of ARS are distinguished: super-acute, very severe, severe, and light.

A *severe form* of ARS has the following symptoms: diarrhea, vomiting, fever, bleeding from skin and mucous membranes, internal hemorrhage, reduction in the number of leukocytes and other white blood cells (lymphocytes and trombocytes).

A *light form* of ARS consists of several well-defined stages:

1st : prodrome stage, characterized by loss of appetite, nausea, vomiting, diarrhea, salivation, paroxysmal abdominal pains, and dehydration. These symptoms appear in minutes, lasting approximately two days.

2nd : asymptomatic stage, characterized by disappearance of the above-mentioned symptoms. This stage lasts from several days to one week.

3rd : stage during which symptoms recur and signs of significant bone marrow injury appear (hemorrhage, anemia, etc.).

4th : recovery stage, lasting from several weeks to one month.

After the accident on the Chernobyl nuclear power plant, from 1987 to 1998, 10 people died from radiation exposure in the range of 1.3-5.2 GY. Among these victims, three perished from cardiac failure, two from significant injuries of the bone marrow, two from liver cirrhosis, one from pulmonary necrosis and tuberculosis, and one from acute myeloid leukemia. Heavy radiation burns and cataracts were observed in 56 victims of the accident. Almost all patients afflicted with a light case of ARS developed functional impotence; of these, 14 victims were able to give birth to healthy children.

In recent years, radiation skin burns have been designated by the term "skin radiation syndrome".

This syndrome more frequently affects injured areas of the skin. Intact areas become involved only with large doses of radiation. Skin radiation syndrome is characterized by redness, dryness, and formation of surface defects. These changes may improve, yet may also potentially recur over time. Areas of radiation burns may subsequently develop pigmentation changes, hardening, or ulcerations.

There are five stages of skin radiation syndrome.

1. Prodrome, there is a 24-72 hours.
2. Manifest, symptoms after days, four weeks.
3. Sub acute develops after four to six weeks.
4. Chronic occurs three months to two years.
5. Later seen in decades.

[n people, providing first aid in the irradiated zone of Chernobyl accident (doctors, nurses and others), we observed the so-called contact skin burns (redness, itching, peeling).

The famous physicist A. Becquerel arose burn the skin, after the fact as in the side pocket of his jacket was a test tube with radioactive element. Radiological burn distinction from thermal, chemical. They not appear immediately.

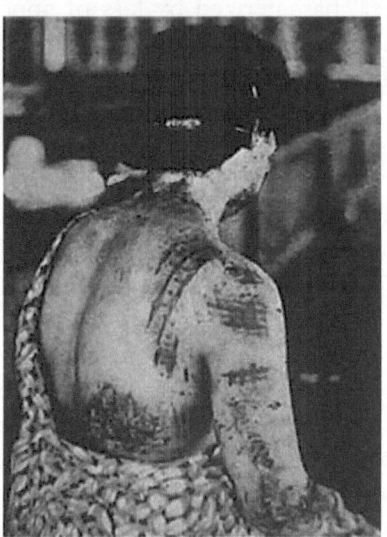

Thermal and radiation burn

After the nuclear bombing in Japan, scientists noted for the first time that men were more susceptible to the effects of radiation than women. They appeared to develop ARS at much lower exposure levels. This was explained by hormonal differences as well as by the degree of reaction to stress. Such data failed to coincide with observations of Chernobyl victims, in whom women were found to be more susceptible to the effects of radiation exposure.

In Japan, bleeding from skin, mucous membranes, and within internal organs was found to be one of the principal manifestations of ARS. This occurred primarily as a result of bone marrow injury, which caused a quantitative reduction in coagulation factors and thrombocytes (clotting cells). Ensuing disturbance of blood coagulation resulted in bleeding, which became the principal cause of victims' deaths. In addition to radiation exposure, an enormous blast wave as well as high ambient temperatures may have concomitantly caused injuries in victims of nuclear bombings in Japan. As a result of this possibility, some experts doubt that all human injuries occurred exclusively from radiation exposure.

Diagnosis of acute radiation sickness does not present significant difficulties in a setting of confirmed radiation exposure. Details about the exposure level represent critical information and determine methods of treatment and prognosis. Any suspicion of ARS necessitates routine blood monitoring for erythrocyte, thrombocyte, and leukocyte content. If possible, cytogenetic analysis of blood should be performed in order to evaluate for chromosomal aberrations. If no suspicion of radiation exposure exists, other illnesses manifesting analogous symptoms should be excluded (severe infection, etc.).

In addition to some of the common signs of ARS, other symptoms of radiation injury arise based on the predilection of radioisotopes for various organs (gastrointestinal, central nervous system, etc.). This targeted preference primarily depends on a chemical composition of these radioisotopes (**Table 1**).

As previously mentioned, children are more susceptible to the effects of radiation when compared to adults. Therefore, signs and symptoms of ARS develop at lower levels of radiation exposure in children.

Information presented above attests that some of the features of ARS also depend on the source of radiation exposure (nuclear weapons, accidents, etc.).

Late Radiation Injuries

The definition of "late radiation injuries" is not limited by specific time periods. These injuries may take place months, years, and even decades following radiation exposure. Numerous conditions determine the onset of clinical symptoms. These include the radiation dose, duration and type of exposure, radioisotope half-life, affected organs, preceding health status, timing and quality of treatment and prophylaxis, social status, etc. Therefore, classification of late radiation injuries presents certain difficulties.

The most common late manifestation of radiation exposure includes various types of cancer as well as congenital and genetic disorders. These arise directly from ionizing mechanisms of radiation, which may also induce other types of injury (targeting individual organs, biological processes, or whole systems of the organism).

Recently, Japanese authors published interesting data regarding specific types of injuries arising from radiation exposure. These were based on long-term follow-up and observation of the nuclear bomb victims and their descendants.

Radiation and cancer

A relationship between various forms of cancer and radiation exposure was first noted at the dawn of radiation era. For instance, cases of skin cancer were observed in professionals who came in routine contact with radiation (physicians, technicians in radiology departments, and others). Subsequent numerous experiments convincingly confirmed this connection. However, with improvement of radiation protection in the workplace, occurrences of cancer became much less common.

Over time, contemporary techniques of molecular biology and biochemistry have gradually substituted long-standing epidemiological studies that established the link between radiation and cancer. Application of such methods has allowed to prove scientifically the role of radiation in the development of cancer.

In areas contaminated with radiation, a distinct possibility exists that other factors may also facilitate the development of malignant tumors (for instance, certain chemical substances). In response to this possibility, several methods have been developed to confirm radiation-based origin of cancer. These include biological estimations of the radiation dose and evaluation for the presence of certain microscopic characteristics of cellular radiation injury.

Increased incidence of leukemia and other malignant tumors has been detected in children born to mothers with a history of radiation exposure. According to the data from a 10-year follow-up of 15 million children and 350 thousand twins subjected to intrauterine radiation exposure, the risk of leukemia and solid tumors increased significantly (children—1.5 times, twins—2.2 and 1.6 times respectively).

Radiation catastrophes of the previous century prompted a thorough evaluation of this multifaceted problem. From an oncologic perspective, increased prevalence of various types of cancer was noted after the nuclear bombing in Japan. Japanese authors currently believe that cancer represents one of the most severe consequences of radiation exposure. Its incidence varies across various organs of the body. Cancers of the thyroid gland and breast represent the most commonly observed types. They are diagnosed in 10 out of 1,000 people exposed to radiation, namely in 1% of cases. Cancers of other organs arise in one person out of 10,000 irradiated, or in 0.1% of cases.

Cancer became the principal cause of death in years subsequent to nuclear bombings in Japan. Of note, medical statistics methods underwent a significant change since 1940s. This, undoubtedly, influenced reported numbers of induced illnesses, particularly cancer. Despite this, general cancer patterns and statistics have not changed appreciably over time.

Long-term research at the end of the twentieth century achieved an enormous task of enumerating and analyzing consequences of radiation exposure on human health. A significant increase in prevalence was established for the following types of cancer: leukemia (with the exception of lymphoid and T—cellular subtypes), breast, thyroid, large intestine, stomach, lungs, and ovaries. To a lesser degree, increased prevalence was also noted for cancers of the esophagus, salivary glands, liver, skin, bladder, nervous system, and hematopoietic organs (myeloma and malignant lymphoma). Further studies are underway evaluating changes in prevalence of chronic lymphoid leukemia, cancers of the pancreas, gall bladder, rectum, uterus, and bones.

Not surprisingly, people exposed to radiation during childhood or adolescence develop cancer more frequently during their lifetimes. The incidence and prevalence curves of various cancer types vary with age. In addition, "cancer peaks" vary with elapsed time after radiation exposure.

After the Chernobyl accident, from 1986 to 2000, the incidence of bladder cancer increased in Ukraine from 26.2 to 43.3 per 100,000 people. It was subsequently discovered that urine excretes approximately 80% of cesium taken up by the organism. This in turn explained a rise in the incidence of bladder cancer. On the basis of these thorough studies, conducted together by Ukrainian and Japanese researchers, a hypothesis was proposed regarding the role of free radicals in carcinogenesis of bladder cancer.

In recent years, the average annual increase of different types of cancer was 5.5% among irradiated people, compared to 1.5% in the control group (consisting of those not exposed to radiation). The highest index of average annual cancer rate increase -7.9%—was observed in cleanup workers and inhabitants living in the 30-kilometer (19 mils) zone around the site of Chernobyl's accident. Cancer of the colon, kidney, and bladder was observed most frequently.

Numerous scientists predict that history of radiation exposure will continue to remain a significant causative factor in the development of cancer. As a result, routine observation and follow-up of radiation victims and their offspring must continue for a long time. Various formulas have been developed to calculate the risk of cancer in people exposed to radiation. Their validity and reliability continues to be thoroughly checked and evaluated.

Analogous cancer data has been acquired by studies evaluating consequences of other nuclear plant accidents. The accident on the atomic enterprise "Mayak" (city of Kishtim in the Sverdlovsk region in the former Soviet Union) for many years has been directly responsible for increased incidence of several cancers, specifically of the breast, lungs, blood, brain, genital organs, and bladder.

The timing of cancer appearance in victims of radiation exposure varies depending on a setting (nuclear bombing, nuclear power plant accidents, etc.). An association has been established between certain cancer types and patient's age. For instance, breast cancer predominantly strikes women 40-45 years of age, while thyroid cancer mainly occurs in children. In addition, cancer may arise many years following radiation exposure. Unfortunately, predisposition to cancer may be transmitted to the victims' offspring. It has been predicted that cancer will comprise 10% of all illnesses in cleanup workers of nuclear power plant accidents.

Diseases of the thyroid gland, specifically thyroid cancer, have formed a new focus in the field of nuclear medicine. During the initial stages following Chernobyl's accident, iodine[131]—which accumulates in the thyroid gland—dominated the radionuclide spectrum. Therefore, the thyroid gland became the first health target of Chernobyl's fallout.

Thyroid gland is an unpaired organ weighing 20-30 grams, located in the neck in the region of thyroid cartilage. A healthy thyroid gland is invisible; however, it can be palpated in people with relatively small amounts of subcutaneous fat. Thyroid gland produces hormones that participate in numerous processes essential for normal functioning of the organism.

In the former Soviet Union, thyroid cancer was encountered extremely rarely prior to Chernobyl's catastrophe, and was practically never observed in children. This data coincided with publications and reports from Europe and America. Because of the rare incidence of thyroid cancer, its diagnosis, prophylaxis, and treatment were not extensively evaluated.

Studies conducted following nuclear bombings revealed that cancer of the thyroid gland appeared approximately 6-7 years after exposure, and this predisposition remained for approximately 20-30 years. A linear growth of thyroid cancer incidence with respect to the radiation dose was also noted. However, inverse relationship was identified with very high doses of radiation.

Initial cases of thyroid cancer occurred 3-4 years following Chernobyl's catastrophe. This was a shorter time frame when compared to nuclear bombings in Japan. Prevalence of thyroid cancer in children increased **50-100-200** times in Belorussia and **5-7-10** times in Ukraine. In Russia, a smaller increase in the prevalence of thyroid cancer was noted in several contaminated regions. Children from 5 to 14 years of age, as well as those exposed to radiation from birth until 4, formed the predominant group among all patients with thyroid cancer. For all age periods, women developed this disease more commonly.

Elevated rates of thyroid cancer in children occurred secondary to high sensitivity of the thyroid gland to radioactive iodine. In Chernobyl, this substance was delivered primarily through consumption of dairy products. When compared to adults, similar radiation levels induce thyroid disease with more frequency in children. This is due to a higher degree of accumulation of radioactive iodine in thyroid glands of children and elevated sensitivity of the gland to its effects. A possibility of underlying genetic predisposition in children has been proposed to explain this phenomenon.

An increase in prevalence of thyroid cancer in adults and children started in 1990 and continued to grow until 1998. Rates subsequently stabilized in children, at a higher level than prior to the accident. Experts predict that another surge of thyroid cancer may occur in 10 or more years, with the risk of this disease remaining for over 50 years.

Among adults, especially in cleanup workers of radioactive contamination, elevated rates of thyroid cancer occurred later and thus far have shown a tendency for increase. However, according to latest conclusions by experts in the field (2004), increased prevalence of thyroid cancer was highly pertinent for only a relatively short period of time after the Chernobyl accident.

A later onset (more than 11 years) of thyroid cancer was recorded in adults after the hydrogen bomb testing on Marshal Islands. Observations in other locations of radiation accidents in the U.S., specifically those occurring in the 1940-60s, failed to establish a clear association between thyroid cancer and radiation exposure. This may have been related to a different pathway of radioisotope entry into the organism.

Victims of the Chernobyl's catastrophe acquired radioactive iodine[131] in three different ways—through skin and organs of the respiratory and digestive tracts. In the U.S., radiation from accidents was better contained, thereby limiting entry of radioisotopes through respiratory and digestive organs.

Studies of atomic bomb survivors in Japan reveal a linear relationship of the incidence of thyroid cancer with age and radiation dose. However, very high doses of radiation reverse this correlation.

Microscopic and molecular analyses of radiation-induced thyroid cancer have established that affected patients predominantly develop a more aggressive (proliferative) form of this tumor. After a long-standing discourse, radioactive iodine has been accepted as the primary cause of thyroid cancer.

So what are some of the signs of thyroid cancer?

The initial stages of this disease are usually marked by absence of noticeable manifestations. Signs of hypothyroidism or hyperthyroidism appear relatively infrequently. Enlargement of the thyroid gland may be noted with visual inspection or by palpation, both of which are facilitated by the gland's fairly superficial location. Adults may examine their thyroid glands independently (self-monitoring), while children require assistance (from parents, medical staff, etc.).

It has been noted that radiation-induced thyroid cancer has a propensity for rapid growth and metastatic spread, which is defined by propagation to other organs (lungs, lymph nodes, liver, etc.). Therefore, diagnosis may be initially established on the basis of metastatic signs of cancer, or in the advanced stages of the disease. Typical signs at that point may include blood—spitting, enlarging lymph nodes above the clavicles and in the armpits, bone pain, and fever.

Contemporary diagnosis of thyroid cancer consists of neck palpation in the region of the thyroid gland, ultrasound and radioisotope scanning, and biopsy of the gland. Certain immunological tests may also be helpful in this setting.

Physical examination of the thyroid gland plays a significant role especially in the beginning of the disease process, during which symptoms and signs are frequently absent.

Thyroid palpation is performed with a patient in a sitting position. During swallowing, the thyroid cartilage, trachea, and intimately associated thyroid gland displace upwards, facilitating palpation of this organ. Self-monitoring patient or family members may perform this examination routinely. Intervals between examinations (weeks, months, etc.) are individually established. When compared to physical examination, special tests (ultrasound, radioisotope scanning, and especially a biopsy), contribute considerably more information regarding the status of the thyroid gland.

Ultrasound scanning is based on interpretation of acoustic waves, which are reflected by tissues and formatted on screen to depict an image of the scanned organ. Organ tissues generate different signals depending on their density. This potentially enables the ultrasound technique to distinguish benign tumors and cysts from malignant cancer. This technique also allows to identify thyroid growths before they become palpable.

Radioisotope scanning represents a method of visualizing anatomical distribution of an intravenously introduced radioisotope. It allows to obtain information about the size, contour, and density of the thyroid gland.

Biopsy of the thyroid gland enables analysis of cellular material under a microscope. Special syringes are utilized for removal of cells from the gland (aspiration biopsy technique). However, surgical excision of tissue yields a more definitive biopsy result (surgical biopsy).

Diagnostic methods discussed above are indicated for all fallout victims that inhabited and currently live in areas contaminated with radioactive iodine. This cohort carries a distinct risk for the development of thyroid cancer. In this group, a subset of victims should be prioritized for a more aggressive screening. Such victims should include children younger than 14 years of age, cleanup workers, and those with radiation-induced thyroiditis.

Individual dissymmetry data represents critical information in potentially determining the priority of victim follow-up. In children exposed to high radiation doses, the risk of thyroid cancer increases six times when compared to those receiving minimal levels of exposure. Unfortunately, thyroid dissymmetry information remains mostly unavailable. To a certain extent, this deficiency has been compensated by various indirect methods designed to retrospectively estimate the exposure dose.

Radiation-induced reproductive, genetic, congenital, and other illnesses

Disorders of fetal development in a setting of radiation exposure were experimentally discovered a long time ago. Unfortunately, observations of radiation effects also became possible in people, secondary to atomic bombings and large nuclear accidents. These caused significant radiation exposure in many people.

Observations of radiation victims began immediately following the nuclear bombing in Japan and continued to this day. Similar to other nuclear accidents, Chernobyl's catastrophe created a firm foundation for extensive and thorough studies of multifaceted problems arising in victims of radiation exposure.

Research on impact of radiation generally focuses on the following aspects:

* Birth rate;
* Pregnancy and post-partum period;
* Health status of children;
* Maternal gynecologic health;
* Congenital and hereditary diseases induced by radiation exposure during pregnancy;
* Influence of parental radiation exposure on their offspring;

A sharp decline in the birth rate occurred following the atomic power plant accident in Chernobyl. In some regions, this rate was even lower than during the Second World War. To a large degree, this was associated with a concomitant rise in a number of abortions that affected early stages of pregnancy. It was subsequently established that fetal-placental insufficiency, arising as a result of radiation damage, was the main reason for elevated abortion rates. Some reports directly showed placental accumulation of certain radioisotopes. In addition, numbers of induced abortions also increased in the first several years following the accident. This rise was fueled by parental fears regarding potential congenital and birth defects in their offspring.

Women exposed to radiation more frequently developed nephropathy of pregnancy, premature labor, and postpartum complications (hemorrhage, infection of the genital organs, etc.). An increase in endometriosis and gynecologic cancer (ovaries, uterus) was specifically noted. In addition, the incidence of uterine myoma (fibroids) significantly increased after the nuclear bombing.

Radiation-induced impact on the fetus has been evaluated in detail. The negative influence of radiation exposure on the fetus is most prominent during the first 36 hours following conception. Radiation effect is especially severe in the first two weeks of pregnancy and frequently results in fetal loss. Death may occur even prior to finding out about the pregnancy. Even if the fetus survives radiation exposure, various congenital defects are likely to develop. Their severity is directly proportional to the radiation dose.

Numerous severe developmental defects, especially of the brain, may arise between the 2nd and 15th weeks of pregnancy in response to large doses of radiation. From 16 to 26 weeks of pregnancy, only high exposure levels are capable of inducing defects in development. After the 26th week, fetal sensitivity to radiation becomes equivalent to that of a newborn and congenital defects no longer develop at this point. However, increased risk for the development of malignant tumors remains.

Overall, fetal sensitivity to radiation is 10-300 times higher when compared to the adult organism.

The frequency of various fetal defects has been calculated with respect to the timing of radiation exposure. Embryos irradiated in the first two months of development acquire defects in 100% of cases, from three to five months in 64%, and from six to ten in 23%. Fractionated exposure, or recurrent radiation with small doses, leads to more significant damages, since radiation impact falls on many different types of embryonic cells. As a result, injury strikes numerous rudimentary organs in critical stages of development.

Defects of reproduction and fetal development (congenital and genetic diseases) were also noted in other European countries, which experienced elevation in background radiation after the accident on Chernobyl's nuclear power plant (Germany, England, etc.). In England, congenital defects increased by 50%.

According to Israeli scientists, who studied many repatriates from contaminated regions of the former Soviet Union, the number of congenital and hereditary diseases increased seven times. Almost half (45%) of children, born to mothers subjected to radiation during 7-15 weeks of pregnancy, developed signs of mental retardation. Furthermore, microcephaly, growth delays, Down's syndrome, and cardiac defects were observed in children of women exposed to radiation in the first half of pregnancy.

Congenital defects induced by radiation exposure number in dozens. In addition to the ones discussed above, they also include skull shape changes, sunken chest, dislocations, diseased teeth, strabismus, cataracts, retinal disease, glaucoma, cardiac defects, diseased kidneys and genital organs, and others.

On the basis of long-term data gathered from victims of nuclear bombings, significant increases occurred in the incidence of micro cephalic, mental retardation, growth and developmental delays, poor school performance, and low IQ (intelligence quotient).

According to certain researchers, exposure of the embryo to small radiation doses may cause subtle disturbances that would not be detectable by contemporary methods, but which may contribute to the development of defects in the future. Induced changes on a cellular level may also lead to the development of inheritable genetic mutations.

Presented material depicts negative impact of radiation on the reproductive process and on raising the incidence of congenital and genetic disorders. Developmental defects may arise as a result of radiation exposure during the embryonic stage as well as from transmission of parental radiation-induced genetic mutations. As discussed, genetic changes may be subsequently transferred through generations, thereby adversely affecting whole populations.

Long Term Exposure to Low Levels of Radiation

The unprecedented scale of Chernobyl's catastrophe created the urgency to assess all of the negative influences of radiation on human health. In response to this nuclear accident, comprehensive research programs were established to evaluate long term sequels of radiation in practically all organ systems of the body. Accumulated data suggests that radiation effects are not limited to the development of cancer, congenital defects, or genetic abnormalities.

Medico-demographic indices

Presently described radiation-induced illnesses affect not only whole organ systems, but also the individual physiological processes in the human body (discussed below). As such, they directly influence medico-demographic indices, including general indicators of population's health. Widely accepted health indices include birth and mortality rates, population's lifespan, incidence and prevalence of various diseases, and disability rates.

In the aftermath of Chernobyl's radiation fallout, comprehensive studies established that the average population lifespan decreased in contaminated regions. This statistic especially affected the male population. Concomitantly, it was discovered that numerous illnesses increased in prevalence. Moreover, radiation-induced diseases were noted to follow a more complicated course. This further contributed to subsequent increases in mortality rates and the overall reduction in the average lifespan.

According to the data from Ukrainian sources, 10 years after the Chernobyl's accident the average lifespan of men fell below the level found in 20 poorest countries in the world. In the same period, various illnesses occurred in 84% of radiation victims and 92 % of cleanup workers. Presently, Belorussia's mortality rate is twice its birth rate. In addition, worsening of many chronic illnesses was noted. The prevalence of various diseases also increased in children, especially affecting the neurological, cardiovascular, and musculoskeletal systems.

As previously discussed, radiation exposure was one of the factors that caused a significant decline in reproductive rates (or birth rates). Increased prevalence of diseases affecting the cardiovascular, nervous, and digestive systems secondarily led to increased incidence of individuals with disability. Various health indices especially deteriorated in cleanup workers of contamination, particularly affecting the age group of 40-45 years.

Impact of radiation on whole populations represents a topic of significant interest. Many disorders of human health in a setting of nuclear accidents differ from all other catastrophes in its accompanying genetic injuries, which may transfer through generations. In terms of overall radiation effect, some experts believe that low levels of exposure in a large number of people are equivalent to large levels of exposure in a few individuals. From a population injury point of view, the genetic risk posed to 100 people, receiving a dose 0.01 SV, is equivalent to the risk of 10 people, receiving a dose of 0.1 SV, and to the risk of one person, receiving 1.0 SV.

*   *   *

In order to categorize the multifaceted influence of radiation on people's health, an organ-system approach will be utilized.

The hematopoietic system

The bone marrow and its principal product—circulating blood in peripheral vessels—anatomically define the hematopoietic system. Sensitivity of the hematopoietic system to radiation, which may induce some of its more severe abnormalities (see "Radiation and Cancer"), has been described since the dawn of radiation biology and nuclear medicine. Therefore, bone marrow and blood represent critical targets of routine evaluations following exposure to radiation (from accidents, radiotherapy, professional activity, etc.).

As a rule, acute radiation sickness reduces all functions of the bone marrow and causes severe disturbances in the entire organism (see ARS). Shortly following the accident on CNPP, a number of quantitative and qualitative changes in the cellular composition of blood (erythrocytes, leukocytes, and thrombocytes) was identified in both adults and children. Some changes proved to be transient and were optimistically considered as a compensatory reaction of the organism to radiation.

However, in subsequent years, persistent alterations in erythrocytes and leukocytes were identified in children. This was accompanied by a concomitant reduction in hemoglobin levels. After approximately 10 years, several abnormalities of the bone marrow (dysplasia, aberrant production of blood, and others) also became more prevalent. Moreover, hematopoietic conditions such as aplastic anemia followed a more difficult clinical course.

Increased incidence of hematopoietic diseases is expected to occur in Chernobyl victims during subsequent 20-30 years. Long-term follow-up of radiation victims would provide accurate rate estimations of radiation-induced abnormalities affecting the hematopoietic system.

The immune system

The Nobel Prize winner Paul Ehrlich (1854-1915) was one of the pioneers in the field of immunology (immunity of the organism). During his time, as in subsequent decades, this concept was pertinent only to infectious diseases. The development of theoretical sciences considerably expanded the overall understanding about human immunity.

According to contemporary teachings, the immune system protects the organism from many external (infections, chemical substances, radiation and others) as well as internal factors (cells from different organs, etc.). The immune system ultimately functions to guard the organism against the disease and represents the first line of defense against potentially harmful processes.

Specialized organs perform basic functions of the immune system. These include: bone marrow, spleen, lymph nodes, thymus, dedicated cellular clusters within walls of the small intestine (Peyer's patches), tonsils, adenoids, appendix and lymphatic vessels. White blood cells, referred to as lymphocytes, execute functions of the immune system on a cellular level. Two basic types of lymphocytes exist—T-lymphocytes and B-lymphocytes. They number about one trillion in the blood stream. Furthermore, many additional substances—commonly called mediators—supplement and participate in various immune processes.

As noted above, organs comprising the immune system are very sensitive to radiation. They may sustain injury from exposure levels as low as 100 REM (0.1 GY). Three principal abnormalities generally arise from radiation exposure—immunodeficiency, allergy, and autoimmune reactions.

Immunodeficiency decreases resistance of the organism against infectious diseases, including the severe types (Acquired Immunodeficiency Syndrome, AIDS). In turn, immune deficiency also leads to an increase in disease prevalence. Chernobyl's catastrophe gave rise to a new medical concept designated "Chernobyl's AIDS." This abnormality was observed fairly often and carried serious health consequences.

Allergic disturbances modify immunologic mechanisms, thereby allowing for appearance of a number of related diseases (bronchial asthma, skin reactions, etc.). Autoimmune abnormalities also form the basis for numerous systemic illnesses (lupus, insulin-dependent diabetes, etc.).

Radiation exposure generates complex derangements of both the hum oral (antibody-based) and cellular immune systems. A number of abnormalities affecting the immune system were noticed immediately after the Chernobyl's accident, with some persisting to this date. Considerable attention has been given to cytogenetic changes identified in cleanup workers as well as people inhabiting contaminated regions. These cytogenetic changes were observed as late as 10 years following the accident and directly correlated with the radiation dose.

The immune system plays a significant role in majority of conditions arising form radiation exposure (such as cancer, genetic and endocrine disorders, internal organ injuries, and many others).

The nervous system

The nervous system consists of two interconnected divisions—central and peripheral (vegetative). The brain represents the anatomic organ comprising the central nervous system, while nerve cell bodies (ganglia) and trunks make up the peripheral division. The central nervous system performs all mental processes (thinking, memory, and many other). One of the main functions of the peripheral nervous system lies in regulating the activity of all internal organs (heart, kidney, liver, etc). Both systems constantly interact with one another.

According to classical rules of radiobiology, central nervous system belongs to a group of low-sensitivity organs that are resistant to radiation. Information provided below indicates that this postulate should be carefully reexamined.

Even in the early period after the accident in Chernobyl, diverse psychiatric disturbances were noted to affect victims of radiation fallout. However, they were largely interpreted as manifestations of radio-phobia (fear of radiation). This explanation outwardly emphasized the lack of significance given to psychiatric symptoms.

In the same time period, signs of injury affecting other systems of the organism were explained as disturbances of the peripheral nervous system. As a result, conditions such as "asthenic syndrome" and "vegetative dystonia" arose. These irresponsible diagnoses were given even to the seriously ill hospitalized patients. Of course, these seriously ill patients had a second—and true—diagnosis of radiation exposure. Unfortunately, this information was readily accessible.

Such interpretation of symptoms was necessary for the Soviet and Republic governments in order to conceal the gravity of medical consequences arising from radiation fallout in Chernobyl. For victims of exposure, this caused harmful delays in diagnosis and treatment.

Following the accident, long-term follow-up of victims allowed better characterization of disorders affecting both central and peripheral nervous systems. Comprehensive studies provided sufficient data to consider that brain injury was caused not only by direct exposure to radiation, but also by systemic disturbances arising in irradiated organisms (such as malfunctioning metabolic processes, immune system abnormalities, etc.). Growing evidence of radioisotope accumulation inside the brain supported direct radiation-induced injury of this organ. Specialized assessment of irradiated brain tissue further revealed that cellular injury was accompanied by numerous disturbances in biochemical processes.

Several research studies compared large and small doses of radiation exposure, in order to establish the differences in subsequent brain injury. High levels of radiation appeared to induce the following abnormalities: cerebral form of acute and chronic radiation sickness, inflammatory changes (encephalopathy), and tumors of the brain and surrounding meningitis. Exposure to low levels of radiation may also appeared to cause early and delayed changes in the nervous system. However, congenital defects in the offspring were more likely to arise from chronic exposure to low levels of radiation.

The most frequent abnormalities of the central nervous system included worsening of memory, difficulties with psychological adjustment, emotional disturbances, depression, and interfering phobias. Data is available regarding pathologic personality changes in cleanup workers, including apathy, paranoid thinking, preoccupation with abstract problems, slowness of speech, and reduction in purposeful activities. Seven years after the accident it was also discovered that mothers of small children, living in the immediate proximity of the atomic power plant, suffered some of the most severe abnormalities in mental and emotional health.

Israeli doctors arrived at analogous conclusions while observing immigrants who came from contaminated regions of the former Soviet Union. Poor progress in school and predisposition to psychiatric illness was specifically noted in children. Japanese physicians described similar findings in the aftermath of nuclear bombings in Hiroshima and Nagasaki.

Some victims also developed the psychosocial syndrome, characterized by behavioral issues, sleep disturbance, and headaches. Evaluation of immigrants from contaminated regions commonly revealed psycho-emotional changes, which were expressed in women to a greater degree.

Recently in the United States, a study was completed evaluating mental health in a large group of people that emigrated from regions with increased levels of background radiation (Rose Mary Perez-Foster et al, 2003). It was discovered that even 16 years after the Chernobyl's accident, a fear of possible health consequences still lingers in this population. As previously described, the incidence and severity of radiation-induced health problems inversely correlated to the distance from the CNPP. In other words, shorter distances led to more severe health consequences of radiation exposure.

Specialized tests (electroencephalography) confirmed the organic nature of brain injury and identified permanent lesions within numerous brain structures. The chronic of damages has been maintained by the action of accumulated radioisotopes with long half-lives (cesium, strontium, etc.). In certain cases, electroencephalography may identify central nervous system lesions prior to appearance of clinical signs.

A number of suicides increased among victims of radiation exposure, especially in cleanup workers of contaminated regions.

Various functional disturbances of the peripheral nervous system frequently coincided with anomalies in the central nervous system. Abnormalities in cardiovascular regulation formed the most common manifestation of the peripheral nervous system malfunction, resulting in altered heart rate and rhythm, fluctuation of arterial blood pressure, pain in the area of the heart, and changes in electrocardiogram.

Intensive mental health studies in victims of radiation exposure had not been conducted prior to Chernobyl's catastrophe. Nevertheless, comparable psychological manifestations were encountered in victims of other nuclear power plant accidents, nuclear bombings, etc.

Similar to other nuclear catastrophes, the psychological impact of Chernobyl's accident differed from natural disasters (earthquake, flood, etc.) in that it generated a situation of chronic health threat (for years and decades). In addition, the intensity of perception by people living in the former Soviet Union had been negatively conditioned by many previous cataclysms (revolution, World War II, etc.). A similarly negative influence on perception came from criminal policies of the government, aimed at providing deceitful information for the purpose of diminishing the actual scale of the accident. Over time, people simply ceased to believe.

The cardiovascular system

As previously indicated, malfunction of the peripheral nervous system may produce a range of cardiovascular-type symptoms. It is important to remember that severe cardiovascular illnesses may manifest these symptoms as well.

Following Chernobyl's catastrophe, the incidence of coronary artery disease and myocardial infarction increased, while predominantly affecting individuals younger than 45 years of age. Prior to radiation fallout, people in this age group were essentially healthy. In addition, myocardial infarctions in radiation victims were marked by extensive heart damage and failure. It was subsequently shown that myocardium accumulated radioactive strontium. This finding confirmed that radiation played a direct causative role in myocardial injury. Furthermore, microscopic examination of myocardial tissue revealed severe structural abnormalities of coronary vessels as well as heart muscle itself.

Increased prevalence of bran vascular diseases was also noted in Chernobyl. Progressive worsening and high risk of associated complications (myocardial infarction, stroke, etc.) characterized arising conditions (encephalopathy, arterial hypertension, and sclerosis of brain vessels).

Radiation exposure forms a risk factor for arteriosclerosis and subsequent blockage (via clot formation). Unfortunately, in a setting of radiation, abnormalities of the cardiovascular system often cause disability in younger adults. Further rise in the incidence of cardiovascular disease has been predicted for victims of radiation.

The respiratory system

The respiratory system consists of the upper respiratory tract (nose, throat, larynx, trachea, and large bronchi), medium and small bronchi, and lungs (right and left). To a large degree, radiation-induced injury of the respiratory system depends on the elapsed time from the accident.

In the initial weeks after the accident on CNPP, a "*Syndrome* of radiation injury of the upper respiratory tract" was described. Its manifestations included pain and discomfort in the throat and larynx, accompanied by dry and occasionally hacking cough. Subsequently, mucous membranes of the upper respiratory tract developed atrophy (or, less commonly, hyperplasia). Frequently, inflammation marked by thick secretions complicated the clinical picture.

A change in the composition of electrolytes and micro-elements in mucous membranes lining the respiratory tract significantly disrupted its protective function. Subsequently, melanin-producing microbes colonized the respiratory tree. These microbes decreased the resistance of pulmonary tissues to various pathogens, including bacteria.

For an extended period after the accident, lung infections (pneumonias) were marked by a more difficult clinical course. In the first four years, an especially severe form of pneumonia was observed that was poorly responsive to treatment. As such, it was designated as radiation-related.

Protracted radiation exposure of personnel inside the nuclear facility "Mayak" induced a severe lung injury—plutonium pneumo sclerosis—diagnosed in 123 victims. Clean up workers as well as inhabitants of contaminated territories developed predisposition toward upper respiratory as well as pulmonary illnesses.

Microscopic evaluation of cells in respiratory tracts of radiation victims revealed significant injury, which qualified as a risk factor for lung cancer. Both adults and children developed decreased volumes of respiration, placing them at risk for acquiring lung diseases in the future.

In response to radiation exposure, the incidence of respiratory allergies doubled in children. In addition, prevalence of pneumo sclerosis increased. This condition otherwise represents an atypical finding in childhood.

Adults also developed radiation-induced lung fibrosis. Increased incidence of tuberculosis and its complications (cavernous, disseminated, and infectious subtypes; death) required significant attention. According to available predictions, increased prevalence of tuberculosis will affect all age groups in the future.

The gastrointestinal system

Radiation sources of external as well as internal (through food and water) nature may induce injury to the digestive system. As mentioned elsewhere, mucous lining of the digestive tract (stomach and bowels) is highly sensitive to effects of radiation. This results in increased susceptibility to injury and subsequent development of various ailments.

In a setting of radiation exposure, acute ulcerations and erosions are commonly observed. The internal lining of the stomach frequently becomes atrophic, thereby lowering formation of acidic gastric juice. This results in indigestion and regurgitation of food into the esophagus, producing the reflux syndrome. The course of ulcer disease is further complicated by high prevalence of concomitant dyspeptic syndrome (nausea, regurgitation, diminished appetite), often without accompanying pains. This leads to delays in diagnosis.

Overall, in adults and children exposed to radiation the prevalence of diseases affecting the stomach increases 1.5-2 times. This coincides with an elevated incidence of cholecystitis and cholangitis.

The liver develops functional and structural deficits at later periods after radiation exposure. Liver injury is designated as radiation hepatitis, which carries a poor prognosis.

The endocrine (hormonal) system

The endocrine system (or hormonal system) consists of organs responsible for production of specialized substances referred to as hormones. Numerous organs generate hormones in the organism (thyroid gland, adrenal glands, pancreas, etc.). Hormones participate in diverse processes necessary to sustain the life of an organism, both under normal conditions and in a setting of injury (in particular, radiation). Hormones secreted by endocrine glands ultimately interact with several sites within the brain (hypothalamus, hypotheses', etc.).

Radiation fallout from Chernobyl's catastrophe exerted various influences practically on all functional aspects of the endocrine system. In the initial period after the accident, beneficial adaptive changes increasing the stability of the organism to radiation were discovered. Undoubtedly, this compensatory reaction in the endocrine system was a positive factor.

On the other hand, influences of radiation on mechanisms not participating in processes of adaptation, led to disturbances in interaction between central and peripheral hormonal regulatory components. As a result, significant functional injuries affected activity of numerous systems in the organism. For instance, a reduction in the synthesis of sex hormones in men led to abnormalities in sperm formation and ejaculation. Diminished levels of antioxidants further induced spermatic injury, predisposing to the development of congenital deformities in the offspring.

Radiation-induced (autoimmune-type) thyreoiditis was a rare condition prior to radiation fallout in Chernobyl. However, it was diagnosed especially frequently in the first weeks and months after the accident. Individual cases continued to occur during subsequent years, mainly affecting people living in contaminated regions. It has been predicted that victims of radiation exposure will carry the risk of thyreoiditis for many years.

Clinically, thyreoiditis causes gland enlargement, with a concomitant decrease or increase in hormonal function. In a setting of reduced thyroid function (hypothyroidism), patients manifest generalized weakness, swelling (edema) of the face and extremities, reduction in body temperature, and apathy. Elevated thyroid function (hyperthyroidism) is characterized by palpitations, heart rhythm disturbances, weight loss, excessive perspiration, increase in body temperature, and irritability.

According to the data from Israel, 43% of children repatriated from contaminated regions manifested various functional disturbances of the thyroid gland. Japanese researchers also noted increased incidence of hypothyroidism in the offspring of individuals that were exposed to radiation from nuclear bombings. Both adults and children manifested abnormalities in the endocrine system. In children, functional endocrine disturbances led to delays in physical development.

Radiation thyreoiditis represents a risk factor for the development of thyroid cancer. The incidence of various thyroid gland disorders will continue to increase in subsequent years.

Diseases affecting organs of the endocrine system increased 9-28 times in various geographic regions affected by radiation fallout.

The urinary system

Formation and excretion of urine represents the principal function of the kidneys. This activity significantly interacts with other critical systems in the organism (cardiovascular, endocrine, etc.). Kidneys play an especially important role in excretion of exogenous substances, or those introduced from outside (for example, certain radioisotopes). Moreover, kidneys participate in the regulation of arterial blood pressure, hematopoietic (blood formation), and substance exchange.

Given these numerous functions, it is not surprising that radiation negatively affects the activity of kidneys. In fact, a clinical entity of radiation nephritis has been described for a long time. It denotes inflammation of the kidneys arising in response to radiation therapy aimed at organs situated in the pelvis. This inflammatory reaction differs from radiation-induced nephritis in a setting of nuclear fallout in terms of timing of its appearance, occurring approximately 15-20 years after radiation exposure. This delay in onset also marks another radiotherapy complication—the development of nephrosclerosis (progressive scarring of the kidneys) with subsequent renal insufficiency and hypertension.

Radiation nephritis appeared much earlier (5-10 years) in victims of the Chernobyl's catastrophe. It is necessary to emphasize that radiation-induced abnormalities of the urinary system may not manifest symptoms for a long time. Therefore, their diagnosis requires proactive measures such as routine urinalysis, blood assessment, and monitoring of arterial blood pressure.

Radiation-based derangements of metabolic processes, in particular those causing oxidation of lipid membranes in cells lining nephritic ducts, represent a risk factor for interstitial nephritis in both adults and children. Furthermore, these processes contribute to formation of stones in kidneys and urinary tracts.

Exchange of substances

Long-standing observations of radiation victims revealed a number of abnormalities affecting principal substances requiring constant exchange—namely fats, proteins, carbohydrates, electrolytes, chemical elements, and vitamins. Disturbances of fat exchange were noted to be especially significant. Moreover, activation of damaging oxidative processes was experimentally identified. These processes generate prerequisite lesions necessary for subsequent development of severe illnesses, such as arteriosclerosis and cancer.

Following radiation exposure, both adults and children may suffer from abnormal substance exchange. Disturbances in protein turnover lead to shortage of sulfur-containing amino acids, causing numerous disorders in the organism. This is accompanied by increased levels of protein in blood and alterations in its relative fractions. Anomalous fat exchange represents a risk factor for premature aging of the organism and arteriosclerosis.

Abnormalities in carbohydrate exchange, in particular caused by insufficient production of insulin in the pancreas, cause elevation in blood glucose levels. All these derangements exert negative influences on other forms of substance exchange and on functions of other organs.

Defective maintenance of electrolytes (potassium, sodium, calcium and other) leads to alteration of their levels in blood, with a concomitant increase in erythrocyte content. All of the above-mentioned abnormalities are marked by relative stability over time and principally arise from disturbances in the hormonal system.

Victims of radiation exposure also suffer from abnormal turnover of chemical elements (iron, zinc, cobalt, etc.) This is mainly caused by malfunction of the endocrine system. Moreover, the iodine content in blood increases almost four times in radiation victims. Accompanying abnormalities in vitamin exchange generally result in vitamin A, C, E, and beta-carotene deficiency.

Sensory Organs

Hearing and balance

Hearing loss has been identified in victims of radiation exposure. It is commonly detected with the aid of a specialized instrument called audiometer. Researchers from European countries presently consider radiation exposure as a new cause of hearing loss in the current century. It has been established that radiation adversely affects all aspects of hearing. Hearing loss in radiation victims has also been detected with electroencephalograms.

Frequently, hearing loss is accompanied by disturbances in the vestibular (balance) apparatus, occasionally marked by severe episodes of spinning (vertigo). Vestibular imbalance may also occur without hearing loss. Hearing and balance abnormalities have been identified in 70-79% of cleanup workers.

Vision

Victims of radiation exposure may develop radiation-induced glaucoma (pressure increase inside the eye) as well as cataracts (opacity of the eye lens). Radiation cataracts differ from those related to advanced age in that lens opacity develops from the periphery, and not from the center. Furthermore, clinical course of these conditions tends to be more complicated.

After nuclear bombings, cataracts occurred at varying time points, generally from three months to 10 years. They appeared more frequently in victims that resided closer to the epicenter of explosion. Refraction abnormalities and retinal disease were also noted.

In Austria, where a rise in background radiation was low, congenital diseases of the eyes were identified in children that were exposed to this elevated radiation in uterus.

In the early stages after Chernobyl's accident, the majority of cleanup workers developed changes in the anterior chamber of their eyes, resembling ultraviolet burns. In contrast, children developed changes in the posterior chamber of their eyes, analogous to those found in survivors of nuclear bombings. In subsequent years, an increase in age-related cataracts and their appearance at a relatively young age were noted as well. The same timing of injury was observed for the retina and its blood vessels.

Skin and its appendages

Skin

Prolonged exposure to small radiation doses leads to injuries that mainly affect hands, feet, and lower third of shins (i.e. exposed surfaces). In severe cases, skin injuries may involve 30% of the body surface. Skin changes are characterized by vasodilatation, thickening and growth of skin elements (keratosis), formation of surface ulcers, and appearance of telangiectasias (vascular "stars"). Abnormality in pigment exchange forms a risk factor for the development of melanoma—a malignant skin tumor.

Scarring and kelloid formation represent some of the more severe skin changes associated with radiation exposure.

Nails

Bleeding, pigmentation changes (discoloration), and scarring may occur in the area of the nail bed. These changes usually accompany radiation injuries in other organs.

Hair

High content of radioactive plutonium was discovered in hair of radiation victims. In addition, a mutation that could potentially lead to baldness was identified.

Teeth

Radioactive strontium accumulation was identified in teeth of radiation victims. Its level was especially elevated in men, which was attributed to their significant participation in radioactive cleanup efforts. American authors proposed a method for determining the content of strontium in teeth in order to ascertain the degree of radiation exposure.

Accumulated follow-up data also documents a rising incidence of dental disease in children.

Skeletal system

Radiation cleanup workers sustained the osteopenic syndrome, marked by pain in the lumbar region. Specialized X-ray studies revealed decreased bone density, accompanied by small fractures in vertebral bodies, osteochondrosis, and spondylosis.

Deposits of radioactive cesium have been identified in bones (more frequently in the trabecular subtype). Case reports exist detailing the development of malignant tumors (osteosarcoma) in unusual locations.

Presented findings and observations from victims of radiation exposure convincingly attest to the variety of radiation-induced injuries as well as unique nature of their clinical course.

Treatment of Radiation Injuries and Diseases

History of treatment of radiation damage count more then 100 year.

During this period, significant changes have occurred in many areas of science and medical practice. Only in the last 50 years, treatment of radiation damage, as well as other aspects of radio medicine, came to the state level.

Already today there are some advances in the treatment of radiation injuries, based on the achievements of many sciences. Nevertheless, treatment should be improved. Treatment of radiation injuries and diseases is multifaceted and very complex problem. In this issue there are two basic directions.

The *first direction* involves the use of methods of disposal of radioactive substances.

The *second direction* is treating the consequences of radiation exposure and the resulting diseases.

Disposal of radioactive substances is central radio active substances and reduce their negative effects on the body identified the emergence of a new section radio medicine—*radioprotection*. Substances neutralizing isotopes and their effects on the body, called *radio protectors*.

It is quite obvious that the content of these directions is determined by all the characteristics of radiation exposure (dose, pathways of the isotope, and much more), as well as social-economic indicators of the country.

Such studies on specific programs are conducted by large specialists in specialized medical institutions.

For the first time in 1949 as radio protector used Sodium cyanide and Cysteine.

Since 60—of the 20 century, increasing attention is drawn so coled biological radio protectors. These are substances of natural origin with different properties (adaptogenic, antioxidant, hamo and immune stimulation, antimutogens and others). These properties radio protectors directed many pathological processes occurring in radioactive forcing.

During the Chernobyl accident radio protectors not used. Only the pilots of the helicopters took part in the liquidation of the accident, received the drug domestically produced "Indralin". Judge its effectiveness is not possible because of the lack of publications on this topic.

After the atomic bombing of Japanese cities, radio protectors have been used in patients with human beings.

The Chernobyl accident was a new impetus for the expansion and deepening the development of radio protectors. Intensification of such research also contributes to the likelihood of terrorist use of "dirty bomb".

History radio protectors largely depends on the level of development in many fields of science (radio physics, radioimmunology, pharmacology, medicine and other).

Before the 40—of the past century radio protectors study was conducted mainly in animal experiments. As a result, obtained some positive results.

There are three groups of substances and medications that have radio protective.

1. Decorporantes—substances that accelerate the removal of radionuclide's. They are readily soluble in water and quickly extracted from the body, mainly in the urine. Decorporantes used by mouth or intravenously. These substances are effective in the early periods after falling radionuclide's in the body. In large doses, they are effectively especially for isotopes retained in the bones (potassium, strontium, and others). Most Discorporate little specific and can rid the body of useful substances (vitamins, trace elements, etc.). Therefore, the duration of their use should not exceed a few days. I n recent years drawn attention to decorporantes vegetable origin. Their positive feature is the low toxicity, and therefore the possibility of their use for a long time.

2. Intestinal sorbents, is substances capable of absorbing radionuclide's and remove them from the body with feces and urine. Each sorbent is specific, is absorb one or two substances. Therefore being made to develop integrated Enter sorbents. They are used not only for therapeutic but also preventive. It is very important that the intestinal sorbents, even with prolonged use, do not display useful for the organism. Promising the creation of drugs and food additives not only blocking the absorption of radionuclide's, but also facilitate their removal from the body.

   The most common enter sorbent include stone and charcoal, high-purity cellulose, silica, organic gale drugs. It is noted that the removal of radioactive substances from the body contributes to a sauna, Jacuzzi and other hydro prosecute.

3. Modifiers of radiation injury. These factors include the physical and chemical nature, positively alter the body's response to irradiation. The physical modifiers include temperature, light, electromagnetic radiation. Especially powerful is the impact ultraviolet and electromagnetic fields of ultrahigh frequency. Modifiers affect the different levels of biological organization—from molecular to organism. By modifiers of radiation effects should include the recently proposed by American scientists method of hyperbaric oxygenation. On the basis of modern ideas about the mechanisms of radiation exposure, its use is entirely appropriate. The authors have identified a positive experience with the method of radiation damage in patients receiving intensity is very radiotherapy.

Some therapeutic effect if ingested radioactive iodine has potassium iodide. Adults encouraged to receive one tablet (130 mg) a day for two weeks. Children dose is reduced according to their age. The therapeutic effect of the drug is reduced to the fact that decreases the accumulation of radioactive iodine in the thyroid gland. The effect of treatment is increased only when used in the early periods after irradiation. A much more efficient method of potassium iodide to prevent the destruction of thyroid gland.

The mechanism of action of chemical radio protectors involves a change in cellular reactions, resulting in increased resistance to radionuclide's, as well as the functional activity of cell membranes.

Radio protectors derived from thyol, beta-carotene, aromatic compounds and other substances. Depending on the duration of action, they are used for treatment and prophylactics. Same time radiation terrorism has intensified the development of high standards, new methods of treatment. Over the past few years new classes of substances for the treatment of radiation injuries, they proposed a systematization.

There are four varieties radio protectors.

1) Preparations containing izoflovine.
2) Medications that reduce radiation exposure. Prussian Blue is applied through the mouth, 500 mg capsules once a day, or by inhalation. Length admission is determined depending on the dose. The product absorbs in the gut are isotopes, which are then excreted in the feces. The drug can be given to pregnant women and young children. containing zinc and magnesium. Side effects include nausea, vomiting, diarrhea, fever. After discontinuation of the drug side effects are reduced. Effectiveness of treatment is determined on the basis of isotope

excretion in the urine: its increase is regarded positively. Treatment is carried out only in a hospital. Treatment should be undertaken only under medical supervision.

3) The eliminators, compounds conducive to the provision of radio nuclides. All of these drugs are still under certain ciphers and await completion of clinical trials. Recently, the United States allowed for the clinical use of three drugs—Ca-DTPA, Zn-DTPA and Prussian Blue, helps release the body of plutonium, zirconium, and americium. Side effects include nausea, vomiting, diarrhea, administered intravenously or intramuscularly. During treatment, appointed by the dietary supplements containing zinc and magnesium.

   Side effects include nausea, vomiting, diarrhea, fever. After discontinuation of the drug side effects are reduced. Effectiveness of treatment is determined on the basis of isotope excretion in the urine: its increase is regarded positively. Treatment is carried out only in a hospital. Prussian Blue is applied through the mouth, 500 mg capsules once a day, or by inhalation. Length admission is determined depending on the dose. The product absorbs in the gut are isotopes, which are then excreted in the feces. The drug can be given to pregnant women and young children. Treatment should be undertaken only under medical supervision.

4) Blocking radioactive substances (such as potassium iodide).It is believed that more are effectively pre-treatment should be to use biological methods of radiation damage (as defined aberrations in peripheral blood lymphocytes). Radio protectors used in radiotherapy and shows for people who work with radioactive substances.

According to leading experts radio protectors must meet the following qualities:

*   to prevent acute and chronic radiation damage,
*   taken by mouth,
*   spreader quickly in the body,
*   be chemically resistant,
*   do not have side effects.

As radio protector Russia's authors recommend the use of the drug "Actovegin". It can be used in tablets, intravenously and in ointments. The drug improves microcirculation, which are in breach of radiation injuries.

In connection with this proposed and used new antibacterial drugs. Their use should be combined with antifungal drugs.

Specialists warn that sometimes produced and heavily advertised drugs, radio protective properties of which are very questionable.

In the treatment of radiation injuries large place occupied by antibiotics. It was found that antibacterial treatment is often have option. In connection with this proposed and used for new antibiotics. Their use must be combined with antifungal drugs.

To improve the function of bone marrow, along with some previously used drugs (vitamins, iron, etc.), proposing new. They include various versions of erythropoietin—substances to boost activity of the bone marrow. Regulated by the indications for transfusion of whole blood and its components (packed red blood cells, differential counts and platelets, plasma).

A range of interventions aimed at treatment of lesions of the digestive system.

In connection with the Chernobyl disaster emergency urgency has become the creation of drugs for prolonged use under conditions of low doses of radiation. Along with synthetic radio protector, identified substances of natural origin. Many of these substances have a pronounced adaptive amplification action, and contribute to the adaptability of the organism, in particular, to radiation. Recommended use of so-called functional foods that have a favorable effect on the number of violations.

Among the biological protectors the most extensively studied drugs ginseng, Siberian Ginseng, beta-carotene and mussel hydrolyze. Were noted radio protective properties of drugs hips, mountain ash, common cinquefoil, yarrow, sea salt. A positive feature of biological radio protectors is the lack of side-effect it and the possibility of more long-term use through mouth.

Deep studies have established some mechanisms of action of many biological radio protectors. This allows the use of drugs, depending on the nature of the damage. For example, beta-carotene and plantain reduce the damaging effect of radioisotopes in the cell membranes. Preparations hips (Lipohpormin), rowan (Sorvilin) reduce radiation—due to the decrease in the number of leukocytes (leucopenia) and platelets (thrombocytopenia). The drug stimulates the immune system of mussels.

Radiation—due to the decrease in the number of leukocytes (leucopenia) and platelets (thrombocytopenia). The drug stimulates the immune system of mussels. Milfoil has haemostatic effect and stimulates recovery.

Ukrainian authors proposed complexes broths from plants.

The leaves of periwinkle, birch, green stalks of oat, walnut leaves and yarrow in the ratio 1:2:2:2:1. One tablespoon of the mixture is poured a glass of boiling water. Take half cup 3-5 times per day for 30-40 minutes before meals.

Eglantine 15 grams, hawthorn-two-three grams, fruits oblipihi—15 grams. Mixture poured a glass of water, bring to a boil, insisting 1.5 hours. Apply for 150-200 ml 4-5 times a day. The mechanism of action of the proposed plant has not been fully explored

Utilization either of the other radio protector depends on many factors of radiation injury. Therefore, in the acute and severe cases, intravenous displaying means. With long-term effects of small doses, including during the radiation treatment, are recommended non-toxic substances introduced through the mouth. The observations indicate, however, in favor of the feasibility of their use.

Number of biological radio protectors has good prospects due to the introduction of sufficient raw material resources. They are used by themselves, as well as food supplements and dietetic foods (bakery products, beverages, etc.).

The role of the so-called free radicals in the development of radiation damage, determines whether the widespread use of antioxidants. In addition to drugs, is of great importance the use of foods rich in antioxidants, including vitamins.

There are food products contain large quantitative of vitamins.

C : black currants, parsley, strawberries, citrus fruits, tomatoes, potatoes.
B-1: bred of coarse flour, rice, honey, nuts, meat, milk, beans, corn. **B-2:** leaves of vegetables and shrubs, apples, sprouted wheat, milk, chicken eggs.
B-3: liver, chicken eggs, milk.
B-5: fiber plants, meat products, liver.
B-6: yolks of eggs, cabbage, liver, kidney, milk.
B-9: soybeans, wheat germ, peas, lentils, lettuce, cabbage.,tomatoes, spinach, liver, meat, oats, nuts, bread, cheese, bananas, oranges.
B-12: egg yolks, milk, carrots, beets, tomatoes, liver, blubbery, parcels, apricot.
D: liver, butter, vegetable oil, milk, fish oil.
E: sunflower oil.
A: liver, fish, egg yolks, sour cream, milk, carrots, bins, tomato
K: cabbage, spinach, fruits, liver, yolks.

Especially great role of vitamin **C**, providing a positive influence on many pathological processes. (bleeding, infection, etc.). Proposed appoint vitamin C, depending on the specific situation. With radioactive forcing dose should be increased to 5.0 g / day. In the case of heavy damage dose of vitamin C is recommended to increase to 50.0 g (!) a day, and enter it intravenously. Also suggested that a combination of natural food additives (supplements) resented below include sets,

Also suggested that a combination of natural food additives (supplements, including sets presented below. *

Vitamin C (1gr 3-4 times a day), vitamin E (400 international units 3 times a day), coenzyme Q 10 (100 mg 2-3 times a day). * Glutathione (500 mg 2 times a day).

Treatment should be conducted in an integrated and phased manner, taking into account a number of circumstances (sanitary conditions, the characteristic lesions, the leading symptoms, etc.).

For the victims of radiation, recommends special food rations. An increase of the daily intake of protein at 15% and a decrease in fat—30%.* Vitamin A (10 thousand international units) * Beta-carotene (25 thousand international units once a day), zinc (30 mg once a day). * L-glutamine (3-10 grams. per day). "Bromelain" (2500-500 mg between meals).

The following use of large quantities of vegetables and fruits, seafood (sea cabbage, etc.). Food should contain plenty of fresh fruits and vegetables, legumes, coffee, alcohol, sugar and spices.

When preparing a diet should be remembered accumulation isotopes in various foods equally. Thus, cesium and strontium accumulates in milk, meat, potatoes, mushrooms, forest berries.

Treatment should be conducted in an integrated and phased manner, taking into account a number of circumstances (sanitary conditions, the characteristic lesions, the leading symptoms, etc.).

Along with the admission to or other radio protector, as soon as possible to be evacuated from contaminated areas. It is very important sanitary and hygienic processing of irradiated (shower, rinse mucosa, change of clothes and shoes).

Treatment of acute radiation sickness (ARS) should begin as soon as possible and hold in the hospital, better specialized. Patients are placed in separate sterile, equipped monitoring Chamber, under systematic supervision of specially trained medical personnel.

Treatment is carried out comprehensively with the main damage Transfused blood and its individual components (erythrocytes, leukocytes, platelets), introducing large amounts of fluid and electrolytes. Patients are placed in separate sterile, equipped monitors.

Treatment being integrated with the main damage and leading symptoms. Regularly appointed symptomatic treatment (antiemetic, cardiac and vascular drugs, etc.).

Transfused blood and its individual components (erythrocytes, leukocytes, platelets), introducing large amounts of fluid and electrolytes.

Used antibiotics, antifungal drugs. To stimulate hematopoietic prescribe growth hormone, to introduce stem cells. Regularly appointed symptomatic treatment (antiemetic, cardiac and vascular drugs, etc.).

The experience of the Chernobyl disaster, bone marrow transplant did not justify the expected results. In severe cases, not early 2-3 weeks after irradiation, shown stem cell transplantation of bone marrow. In the case of communicable disease prescribed antibiotics (Penicillin, Canamycin, Tiporin, Rafidin). Simultaneously applied antifungal agents.

The commonly used analgesics and antihistamines. Applied drugs that improve liver function.

In severe cases, not early 2-3 weeks after irradiation, shown stem cell transplantation of bone marrow.

It is proposed to use laser light to improve the immune system. In connection with the defeat of the mouth and digestive tract are facing great difficulties in feeding the victims. Recommended candy, chopped cauliflower, egg whites and whipped cream.

Treatment of widespread and deep burns begins with the removal of nonviable tissue. In the following apply all known modern methods of treatment of Burns. Much more difficult to treat burns on the mucous membranes of the mouth, throat and respiratory tract. Recently, progress has been made in the treatment of coetaneous radiation syndrome, and subcutaneous fibrosis. The method depends on the degree of damage. The treatment lasts from several weeks to several months. In patients with subcutaneous radiation fibrosis has been successfully used the drug "Lipsod", used for 12 weeks.

Its side effects are not detected.

When stable keratosese performed surgical treatment, removal of damaged sites, followed by transplantation of healthy skin. Positively evaluated the combination of derivate Methylxanthine with alpha-tocopherol.

Finding new radio protectors continuing.

Submissions clearly demonstrate that treatment of radiation damage must be individualized, consistent and comprehensive.

Prevention of Radiation Injuries and their Consequences

In the conventional understanding of the primary prevention of radiation damage is very limited. This is due to the suddenness of their appearance. Who knows when an accident happens or will to use nuclear weapons.

Historically, major radioactive fallouts have occurred unexpectedly. Since it is nearly impossible to foresee industrial nuclear accidents or deployment of nuclear weapons, radiation protection and immediate treatment of associated injuries has a limited initial scope.

Timely decontamination of affected regions and neutralization of radioactive wastes remains basic to the treatment of radiation injuries, as is protection and safeguarding of all radioactive sources. Immediate treatment of radiation injuries is possible only in certain industrial settings and also in patients actively undergoing radiation therapy. Principal methods of preventing injury during radiologic procedures focus on continuing modernization of instruments and protocols, improvement in effectiveness of personal protection devices, and routine screening of patients as well as service personnel.

Under specific circumstances, certain radio protective agents may be advised. For instance, periodic intake of potassium iodide is recommended for people residing near nuclear power plants. The use of potassium iodide for prophylaxis against thyroid diseases, in particular cancer, is especially indicated immediately following nuclear accidents that expel radioactive iodine. Such information should be provided to the population through appropriate channels.

Recommendations regarding potassium iodide prophylaxis have been established and confirmed by the Department of Health and the Food and Drug Administration (December, 2001). The dose of potassium iodide depends on the age: adults (up to 40 years) take 1 tablet (130 mg), children from 3 to 18 years—half of a tablet (65 mg), children from 1 months to 3 years—quarter of a tablet (32.5 mg), newborns to 1 month—1/8 of a tablet

(16.25 mg). Breastfeeding women must take the adult dose, plus the dose for a child depending on the age.

Disagreement exists on the necessity of prophylaxis with potassium iodide after the age of 40, when the probability of developing thyroid cancer sharply decreases. Potassium iodide intake must be stopped when levels of radioactive iodine[131] normalize in the environment. Potassium iodide is contraindicated in people with diseases of the thyroid gland (hyperthyroidism, goiter and others), or with heightened sensitivity (allergy) to iodine. Potassium iodide pills are readily available in most pharmacies.

The nuclear accident in Chernobyl resulted in elevated background radiation, with partial contribution from iodine[135]. More than 10 million adults and 7 million children were given daily preventive doses of potassium iodide (15 mg). Side effects were not observed with this dose, and prevalence of thyroid cancer practically did not increase.

Unfortunately, iodine prophylaxis was not instituted immediately in all of the affected areas of the former Soviet Union. Potassium iodide pills were in shortage and technical details of its administration were not well known, even to the medical personnel. The total "effect" of iodine prophylaxis came down to generating large numbers of observed allergic reactions.

Soon after the nuclear explosion in Chernobyl, cleanup workers participating in fire extinguishing inside the damaged reactor received prophylaxis against radiation injuries. For this purpose, a Soviet domestic product called "Indralin" was utilized.

Under circumstances of sudden radiation exposure (such as accidents involving nuclear power plants or detonation of "dirty bombs"), it is necessary to use all measures, directed toward diminishing the radiation dose and its time of duration. Therefore, rapid notification has an enormous value, and must include information regarding the exact location and level of radioactive fallout, composition of radioisotopes, and sites of nearest decontamination centers and treatment facilities. In addition, designated special services should be available for patient transport, food and water distribution, and first aid administration. In order to diminish duration of radiation exposure, first aid should be administered as far away as possible from the source of radioactive contamination.

In U.S., organizational aspects of protection from radioactive exposure are maintained at a high level and continue to improve. People and surrounding environments near locations of possible radioactive contamination (nuclear power plants, radioactive waste disposal sites, etc.) undergo routine monitoring.

Some countries, including the U.S., are developing radiation monitoring for public transportation especially the aviation industry) and for busy public areas (subways, stadiums, etc.). At present, the responsibility for this work in our country lies with the Centers for Disease Control and Prevention (CDC). CDC works together with numerous other establishments (local and state departments of health, power engineering, transportation, agriculture, the Food and Drag Administration—FDA, the Federal Bureau of Investigation—FBI, and many others). A schematic lists levels of preparation and various responsibilities of such agencies within the government ladder.

Not surprisingly, containment of the source of radiation occupies the central place in the initial treatment of radiation injuries. In a setting of radiation hazard, population and specialized services, foremost medical, must be constantly kept informed regarding areas of radioactive contamination, levels and composition of released radioisotopes, etc. For this purpose, various media outlets (radio, television, and newspapers), electronic mail, leaflets, lectures, and telephone communications may be utilized. Presently, a device functioning dually as a cell phone and a radiation dosimeter is being developed. Preparation and retraining of medical personnel and other services continues to take place on a regular basis. Government agencies at different levels oversee this important activity, with significant funds allocated for this specific purpose.

One of the basic means of diminishing radiation exposure is the immediate evacuation of victims from a zone of radioactive contamination as well as various health and hygiene measures (shower, haircut, mouth gargling, and exchange of clothing, foot-wear, and bedding, etc.). These steps are very important not only for the victims, but also for other people, since clothing and footwear may spread radioactive contamination. For the purpose of personal prophylaxis and protection, it is recommended to leave the contaminated areas as soon as possible and wear a facemask. After leaving the radioactive zone, it is necessary to take a shower and replace clothing, footwear, bags, etc. (while placing all contaminated belongings into a plastic bag).

Medications should be placed in plastic containers. It is absolutely prohibited to consume water and food from contaminated areas. If radioactive materials are located out in the open areas, people should immediately enter indoors and thoroughly shut all doors, windows, and external fans.

Secondary to various penetrating properties of different types of radiation, simple methods of protection can be used:

Alpha-rays—light material (for example, paper).

Beta-rays—dense cloth.

Gamma-rays—lead.

In addition to recommendations presented above, treatment of complications from radiotherapy also includes a thorough control of dosing and systematic monitoring, directed toward detection of possible side-effects (via blood analysis, immunologic monitoring, and so forth.).

Various multifaceted models are available to determine the exposure dose and predict its possible complications. Researchers from a large American university recently proposed a model, which allows to estimate the degree of radiotherapy-induced damage in lungs of each individual patient. For the treatment of radiation injuries involving the oral cavity.

American experts recommend preparation called "Benzydamine." Those working under conditions of industrial radiation exposure (atomic power plants, submarines, etc.) must strictly and constantly follow various safety guidelines, routinely undergo dissymmetry measurements and screening for initial radiation changes (aberration of leukocytes in peripheral blood and others).

A possibility of "dirty bomb" use by terrorists has necessitated the development of additional rules and recommendations, in particular for the population of our country. It is emphasized that in a setting of radiation emergency, people located indoors must refrain from going outside and must await further instructions. No timetable for how long something like this would last. Therefore, it is necessary to adequately prepare each house for the worst case scenario.

For instance, the state of Israel mandates that special premises be available for the purpose of personal protection. As a result, apartments in newly constructed buildings include specially equipped rooms—the so-called gas-proof shelters. Similarly, special accommodations exist for possible radioactive attacks. These are referred to as domestic shelters (or 'Shelter in your Home') and, for practical reasons, are located in the basement or a centrally located room (both have a minimum number of windows). This location must contain adequate supplies of most essential things, with quantity dependent on the number of inhabitants. The allocated space should account for domestic animals as well.

Recommended supplies include:

Preserved foods (canned products—meat, vegetable, fruit, dried fruit).
Water in the amount of one gallon for a 24 hour period per person.
Baby food.
Change of linen, footwear, and clothing depending on the season.
Paper towels and plastic dishes.
Plastic bags.
Bed linen, (blankets, pillows, etc.)
Battery-operated radio.
Medications.
Powdered Vitamin C.
Toiletries.
Battery—powered flashlights.

Telephone or cellular phone.
Spare glasses, lenses and their cleaning supplies.
Garbage bags.
Animal food.
First-aid kit.
Soap.
Books and games.
Credit cards and money.

Food, water, dishes, and other supplies are gathered to last for at least three days. Once every six months supplies of food, water and medications must be renewed.

In a setting of a nuclear fallout, prior to entering the shelter clothing and footwear must be changed and contaminated items placed outside in plastic bags. It is necessary to turn off air conditioners, place a plastic rug at the door, and tightly shut all doors and windows. Domestic animals should be moved into the shelter as well. Radios must be turned on at all times in order to receive further instructions.

It is well known that radiation does not possess color, smell, or taste. Finding out about a nuclear fallout (from accidents or "dirty bombs") is only possible through official agencies.

So what individual actions are necessary following a nuclear accident? Proposed recommendations include the following:

Do not panic.
Immediately leave contaminated areas on foot.
It is not possible to use public or personal transportation, since it may be contaminated.
It is necessary to enter immediately into the nearest shelter.
As soon as possible remove clothing and footwear, and place everything into a plastic bag.
Take a shower, or wipe the body with moist towels.
Pay attention to all sources of information (telephone, radio, television).
Follow further instructions from agencies in charge.

Special rules exist for the evacuation of kindergartens and schools, located within a radius of 10 miles from the point of the accident. Instructions have also been developed for evacuation of medical, governmental, and a number of other establishments (such as nursing homes).

Secondary measures against radioactive exposure continue to be developed and improved. In Ukraine they include:

* Protection of population and personnel from radioactive effects.
* Shut down of dangerous nuclear sites.
* Measures limiting the spread of radiation.
* Complex monitoring of the population, workers, and the surrounding environment.

All of these methods are united under the heading *"Minimization of Consequences from Radiation."* Obviously, such methods must continue to improve and remain fully accessible to everyone. Regretfully, these measures are not always carried out entirely or on time.

Thorough and systematic control of food product contamination carries especial significance. In Ukraine, permitted contents of cesium and strontium in basic products of nourishment have been established. Testing of radioactive levels in food is possible only in specialized laboratories. Permitted levels in Ukraine differ from those in Belorussia and Russia.

According to recent communications, the quantity of radioisotopes in food products considerably exceeds the established norms. It has been noted that this applies foremost to products generated by the private sector, which does not tightly follow established regulations. One should remember that accumulation of radioactive materials is unequal across different food products. Maximum accumulation has been discovered in mushrooms, dried and fresh berries, milk and dairy products. Therefore, consumption of these products, especially by children, must be limited. Similarly, consumption of food that was not appropriately inspected is contraindicated. Baby food products must contain a minimum amount of radioactive isotopes.

Tight adherence to allowed levels of radiation in food products would expose the population only to the permissible annual radiation dose. It is important to perform routine estimations of radioisotope content in water supplies, air, soil, and construction materials. For each of these, permissible limits of radioactive contamination have been established. Permissible annual radiation limits have also been determined for workers on nuclear plants.

All people arriving from contaminated regions must be clinically evaluated by a number of specialists (internists, neurologists, endocrinologists, and allergists) and undergo radiation measurements.

Late consequences of radiation exposure may arise in varying time frames. Their severity largely depends on the overall dose, duration of exposure, type of radioisotope, victim characteristics, and external conditions. The radiation dose and names of predominant radioisotopes are critical for proper follow-up of radiation victims. Unfortunately, only a small number of victims in the former Soviet Union have such information.

The use of various diagnostic methods depends on the timetable after radiation exposure. Therefore, the entire array of available methods, which allow to either directly or indirectly establish the radiation dose, should be utilized.

Radiation-induced diseases typically follow a prolonged asymptomatic course. Therefore, radiation victims must learn and use self-monitoring techniques, for the purpose of timely diagnosis of several possible conditions (for example, thyroid cancer). Periodic stays in non-contaminated regions and visits to health centers also make a prophylactic impact on the consequences of radiation exposure.

Scientific educational work in all layers of society (through pamphlets, lectures, television, and radio programs) carries an especially significant value. At baseline, it is important for people to understand some of the basics of nuclear medicine. This implies acquiring knowledge about sources of radiation, diagnostic methods, treatment options, and prophylaxis of radiation injuries. Scientific knowledge in these areas continues to expand, and population must actively incorporate this new information.

CONCLUSION

Similar to other important discoveries, X-rays and nuclear fission (splitting of the atom) proved to be both beneficial as well as dangerous for the humanity. On the one hand, nuclear energy has made a significant contribution to the progress in many branches of science and industry. This especially includes successes in nuclear medicine and power engineering. On the other hand, nuclear energy has become the source of deadly weapons of mass destruction.

Today, widespread utilization of radiation occurs in research, medicine, nuclear engineering, and many other areas. During the last decade, various studies have specifically focused on the influence of natural background radiation on people. In the same time frame, the development and application of artificial radiation sources has continued to improve. However, despite numerous efforts, not all associated dangers have been overcome.

Damaging effects of radiation on people and the environment have been known since the discovery of radiation. An enormous collection of facts, testifying about the diverse impact of radiation on people, has been accumulated through years of intensive research by specialists with different backgrounds (physicists, biologists, physicians, and many others).

Consequences of radiation catastrophes may linger for many years and even centuries. This time frame differs from other major disasters (such as earthquakes, floods, fires, etc.) in that. It is worth repeating that artificially created radioisotopes (for example, Americium), which have been increasingly utilized by people, are more aggressive and longer lasting than those of natural origin (for example, uranium). This, understandably, carries more severe consequences.

The consequences of nuclear bombings of Japanese cities Hiroshima and Nagasaki have been characterized as an "endless misfortune." Numerous negative influences have persisted until today, and potential harms are likely to linger for a long time. Chernobyl's nuclear accident, after almost 20 years and for many years to come will remain an ongoing catastrophe. The same could be stated about the Chelyabinsk nuclear enterprise. Unfortunately, similar determinations apply to all cases of radiation fallout from different sources.

Problems associated with radiation exposure, in particular its health impact on people, have elicited a unified response from the entire world. Numerous world-class scientists and various authoritative establishments from developed countries take a large part in trying to resolve various radiation-related issues.

The impact of radiation on the human organism has been intensively studied for more than 60 years. Unfortunately, this task is going to be relayed to many subsequent generations, with the goal of continuing long-term monitoring in victims of radiation exposure and their descendants (from radiotherapy, nuclear power plant accidents, military maneuvers, etc.).

Despite some discrepancies, numerous observations provide very reliable facts linking radiation exposure to the development of various diseases. These include:

*   Thyroid cancer, occurring especially frequently in younger children.
*   Cancer of various other organs (lungs, breast, stomach, and genital).
*   Genetic diseases and the mutations, the frequency and gravity of which is inversely proportional to the stage of pregnancy.
*   Diseases of the bone marrow (leukemia, anemia, etc.).
*   Disturbances of the immune system (allergy, immune deficiency, autoimmune diseases).
*   Increased prevalence of diabetes.
*   Specific changes in the central nervous system.

The World Health Organization has acknowledged the role of radiation in the development of many illnesses, including cancer of various organs, only in 1995—almost 10 years after the nuclear accident in Chernobyl. This decision was based on growing and convincing evidence gathered by scientists in many different countries.

Radiation-induced diseases of the heart, lungs, gastrointestinal tract, endocrine and urinary systems, eyes, skin and its appendages, and other organs have been marked by difficulties in establishing the diagnosis. Many of them have also followed a protracted clinical course, characterized by poor responsive to available treatments. Indices of human health have markedly deteriorated in places affected by radioactive contamination, where population lifespan and birth rates have decreased. This is in stark contrast to increased rates of sickness, mortality, and disability. It is surprising to still hear a viewpoint of certain scientists, sometimes widely circulated by media outlets, about "myths of Chernobyl" and the overall safety following the accident.

Ongoing evolution of ideas about mechanisms underlying radiation injury remains very important. Chernobyl' catastrophe proves this point very clearly. In contrast to other sources of radiation, Chernobyl's fallout for the first time in the history of the world subjected millions of people to long-lasting exposures of low doses of radiation. It has since been shown that prolonged exposure to small doses of radiation is analogous in its impact to isolated large exposures. This is especially true in terms of cancer development, congenital conditions, and hereditary diseases. This fact is of extreme importance, since the number of people affected by small doses many times exceeds those exposed to single large levels of radiation.

Clear association has been established linking damaging effects of radiation to its different properties (dose, duration of exposure, method of entry into the organism). In-depth experimental studies and clinical observations have led to the development of new medications and treatment modalities. Indications for new drugs have been refined over the years, and side effect profiles have been further elucidated. Together with these novel preparations, radio protective properties of certain plants, flowers, and berries have been simultaneously investigated.

Various methods of rehabilitation and social protection of radiation victims play a significant role in their recovery. With the aid from the world community, social protection of victims has improved and numerous centers have been established to rehabilitate both children and adults.

Measures of primary and secondary protection against radiation exposure have been developed and continue to undergo refinement. Protection of various sources of radioactive contamination (cyclotrons, nuclear power plants, and radioactive waste sites) is of paramount significance. Methods of long term follow-up of radiation victims continue to be developed. They vary to a slight degree among individual countries.

Chernobyl's catastrophe has emerged as a platform for international collaboration. With the fall of the Soviet Union, Chernobyl's accident involved not just three republics, but the currently independent countries of Ukraine, Belorussia and Russia. Hopefully, all nuclear catastrophes can serve as a lesson for the entire world and facilitate prevention and management of radiation consequences in the future.

Communities from many different countries as well as highly regarded international organizations (United Nations, World Health Organization, and others) have made a positive impact in helping to resolve numerous problems related to Chernobyl's catastrophe. Unfortunately, continuing radiation exposure of people in contaminated regions remains an ongoing human tragedy.

The opinion of a well-known American scientist John Hoffman represents a fitting description: " . . . to proceed from the worse and do everything so that consequences would be minimal".

The active stance of radiation victims carries a significant value in terms of diminishing the consequences of their exposure. For this, it is necessary to have an understanding about the sources of radiation, diagnostic methods, prophylaxis, and treatment of radiation injuries. In other words, scientific literacy of a wide circle of people, and their active stance undoubtedly would bring success to resolving numerous issues associated with "radiation and health."

Author would be extremely grateful for the wishes as well as critical remarks from readers.

Glossary of Terms

Aberration	violations related to the protection chromosomes
Absorbed dose	energy absorbed by tissues (effective biologic), expressed in Gray (GY)
Adaptation	adjustment, acclimatization
Adaptogenes	group of compounds that adapt the organism to harmful effects of radiation
Adequate	accordance
Adenoma	benign tumor
Allergy	increased sensitivity of an organism to certain substances
Alternative medicine	support treatments (Acupuncture, Physiotherapy Homeopathy and others)
Allergy	increased sensitivity of an organism to certain substances
Alternative Medicine	health care practices (homeopathy, acupuncture, herbal medicines, physical manipulation, etc)
Anatomy	research on structure of matters organism
Antibiotics	antimicrobial medicaments
Angiocholitis (Cholangitis)	inflammation of the bile duct

Antimutogene	substances affecting mutation
Antioxidants	substances decreasing free radical
Aplastic anemia	anemia from malfunction of the bone marrow, which generates red blood cells
APP/NPP	atomic Nuclear power plant
ARS	acute radiation sickness /Acute radiation syndrome/
Atom	the smallest particle capable of participating in chemical reactions, consisting of a nucleus surrounded by electrons; source of nuclear energy
Atomic Bomb	powerful explosive device, containing radioactive materials (uranium, plutonium)
Atomic (Nuclear) Energy	energy resulting from the splitting of the atom (nuclear fission reaction)
Atrophy	decrease in the size of cells in tissues
Audiometer	instrument for determination hearing
Background Radiation	radiation emitted by radioactive isotopes normally found in the environment (air, soil)
"Besporogovaya" Dose	one of the underlying hypotheses explaining harmful effects from small radiation doses. A dose of radiation that exerts dissimilar impact in different individuals.
Biopsy	the removal and microscopic examination of a sample of tissue from a living body for diagnostic purposes

Bone marrow	organ making blood
Carcinogenesis	process in which normal cells are transformed into cancer, mediated by mutations of the genetic material
Catalysator	accelerator biochemical process
Chromosome	structure consisting of DNA (deoxyribonucleic acid) and associated proteins, located in the nucleus of cells
"Chernobyl AIDS"	immune suppression secondary to radiation exposure
CNPP	Chernobyl nuclear power plant
Collective Dose	radiation dose multiplied by the number of exposed victims
Congenital disease	occurring during fetal period
Cyclotron	machine generating radioactive substances
Cystitis	inflammation of the urinary bladder
Cytogenetic Effect	associated with cellular injury
Decontamination	cleanup of radioactive substances (from soil, buildings, etc)
Demography	science of population
DNA (Deoxyribonucleic Acid)	principal component of a cell's nucleus, associated with transmission of genetic information
Dosimeter	a device that measures radiation levels
Dysphasia	improper development of organs

Dysphasia	improper development of organs
Ecology	science about environment
Electrocardiogram (EKG)	graphical recording of the heart's electric cycle
Electroencephalography (ENG)	recording of electrical activity of the brain
Elimination	purging, riddance
Endometrioses	tumor disease of the inner layer of the uterus
Intestinal sorbents	compounds that absorb substances secreted or excreted from blood into the intestine
Erosion	superficial ulceration
Erythrocytes	oxygen carrying cells in the blood
Feto placental insufficiency	circulatory disturbances in the placenta
Free Radicals	highly reactive compounds formed during injury of cells in tissues
Gene	a hereditary unit consisting of a sequence of DNA
Genetic Effects	consequences of radiation exposure on a victim's offspring
Genetics	the branch of biology that deals with heredity and mechanisms of hereditary transmission
Growth Hormone	secreted by the pituitary gland; promotes growth of the body by stimulating cell division
Hepatitis	inflammation of the liver

Homeostasis	constant internal environment organism
Hyperplasia	increase in the size of cells tissue
Hyperbaric oxygenation	oxygenation under power
Hypophisis	gland, located at the base of the brain and controlling the function of various organs
Immunity	inherited, acquired, or induced resistance to a specific compound or pathogen
Immune Suppression	suppression of the immune system functions
Immune Stimulators	increase functions of the immune system
Immune Suppressors	decrease functions of the immune system
Industrial Exposure	radiation exposure from industrial activities
Industrial Increases	associated with industrial accidents of Radiation
Incorporated Isotopes	isotopes bound within cells of various organs
Inhalation	inspiration of various compounds
Interferon	antivirus compound
Ionizing Radiation	high energy radiation (alpha, beta, gamma) capable of producing ionization in substances through which it passes
Ionizing Radiation Types	Alpha, Beta, Gamma, which differ by wavelength
Isoflavones	compounds neutralizing free radicals

Isotopes	atoms with the same atomic number, but with different number of neutrons
Keratosis	grow surface layer of the skin.
Keloid appearance	escalation tissue tishue injuries on the spot.-
Latent period	time between radiation exposure and appearance of symptoms of organism injury
Leucopenia	decrease of leukocytes (white blood cells) in the blood
Liquidators (Chernobyl)	people participating in the cleanup of contamination
Man-made (Artificial)	
Radiation	radiation arising from atomic splitting (nuclear fission reaction)
Markers	specific substances
Mediators	transmitters biological processes
Medical Genetics	branch of medicine dealing with hereditary diseases and their causes
Microcephaly	small head circumference secondary to inadequately developed brain
Mongoloid	changes in facial features, characteristic of or resembling a Mongol
Monitoring	continuous observation
Myocardium	muscular wall of the heart
Nephritis	inflammation of the kidneys

Nephropathy pregnant	kidney disease during pregnant
Nuclear Scanning	visualization of distribution of intravenously injected radioisotopes
Penetrating Radiation	radiation capable of penetrating skin
Pericardium	outer membrane of the heart
Phytotherapy	treatment by grass
Phobia	fear
Placenta	the location of the embryo
Pneumonia	inflammation or infection of the lungs
Pneumosclerosis	formation of scar tissue in the lungs, usually from severe damage by plutonium
Proctitis	inflammation of the rectum or anus
Proliferation	multiplication of cells of body tissues
Prostatitis	inflammation of the prostate gland
Psychotropic Medications	drugs causing an altering effect on perception, emotion, and behavior
Radiation	emission of energy particles or waves by atoms
Radiation Diagnostics	application of radiation technology for diagnostic purposes
Radiation Illnesses	Illnesses arising from effects of radiation
Radiation-Induced	Illnesses caused by radiation
Radiation Oncologist	physician, specializing in treatment of cancer via radiation therapy

Radiation phobia	fear of radiation
Radiation Protection	measures decreasing the radiation impact
Radiation Protectors	substances, decreasing effects of radiation
Radiation Resistance	resistance of an organism to the effects of radiation exposure
Radiation Sensitivity	increased susceptibility to radiation effects
Radiation Therapy	treatment with radioisotopes
Radioactive Waste	remaining unused radioisotopes
Radioactivity	energy arising from decay of atoms
Radioisotope	radioactive isotope of an element, with an unstable nucleus
Radiology	the branch of medical science dealing with the medical use of penetrating radiation in diagnosis and treatment of disease
Rehabilitation	complex activities for restore health
Reconstructed Dose	determination of the radiation dose by indirect methods, at remote time points following the exposure
Rehabilitation	measures targeting health recovery
Repair Enzymes	enzymes facilitating repair of radiation-induced injury
Reproduction	the process of generating offspring
Respiratory System	organs responsible for air exchange
Retrospective	after something happened

Sepsis	bacterial contamination blood
Somatic Effect	effect on organs of the body from radiation
Sonogram	ultrasound evaluation
Stem Cells	responsible for replacing or stimulating cell division in various organs of the body
Syndrome	composing of symptoms
Teratogenic Effect	causes malformations of an embryo or a fetus
Thermography	dynamic detection temperature
Thymus	body immune system, located on neck
Thrombosis	development of a thrombus, leading to blockage in vessels
Thrombocytes (Platelets)	cells participating in blood clotting
Thyroiditis	inflammation of the thyroid gland
UN	United Nations
Units of Radioactivity	Curie (CI), Becquerel (BQ), Gray (GY), Severe (SV), REM, Roentgen
W HO	World Health Organization

www.ingramcontent.com/pod-product-compliance
Lightning Source LLC
Chambersburg PA
CBHW022017170526
45157CB00003B/1268